Bob Miller's Math for the GMAT®

By Bob Miller

Former Lecturer in Mathematics
City College of New York
New York, NY

Research & Education Association

Visit our website at: *www.rea.com*
For GRE updates go to: *www.rea.com/GRE*

Research & Education Association
61 Ethel Road West
Piscataway, New Jersey 08854
E-mail: info@rea.com

BOB MILLER'S
Math for the GMAT

Published 2009

Printed in the United States of America

Library of Congress Control Number 2007943126

ISBN-13: 978-0-7386-0388-9
ISBN-10: 0-7386-0388-0

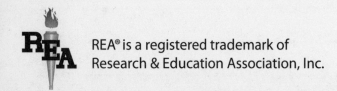
J08

TABLE OF CONTENTS

ACKNOWLEDGMENTS

I have many people to thank.

I thank my wife, Marlene, who makes life worth living, who is truly the wind under my wings.

I thank the rest of my family: children Sheryl and Eric and their spouses Glenn and Wanda (who are also like my children); grandchildren Kira, Evan, Sean, Sarah, and Ethan, my brother Jerry, and my parents, Cele and Lee, and my in-law parents, Edith and Siebeth.

I thank Larry Kling and Michael Reynolds and Mel Friedman for making this book possible.

I thank Martin Levine for making my whole writing career possible.

I have been negligent in thanking my great math teachers of the past. I thank Mr. Douglas Heagle, Mr. Alexander Lasaka, Mr. Joseph Joerg, and Ms. Arloeen Griswold, the best math teacher I ever had, of George W. Hewlett High School; Ms. Helen Bowker of Woodmere Junior High; and Professor Pinchus Mendelssohn and Professor George Bachman of Polytechnic University. The death of Professor Bachman was an extraordinary loss to our country, which produces too few advanced degrees in math. Every year, two or three of Professor Bachman's students would receive a Ph.D. in math, and even more would receive their M.S. in math. In addition, he wrote four books and numerous papers on subjects that had never been written about or had been written so poorly that nobody could understand the material. His teachings and writings were clear and memorable.

As usual, the last three thanks go to three terrific people: a great friend, Gary Pitkofsky; another terrific friend and fellow lecturer, David Schwinger; and my cousin, Keith Robin Ellis, the sharer of our dreams.

Bob Miller

DEDICATION

To my wife, Marlene. I dedicate this book and everything else I ever do to you.
I love you very, very much.

BIOGRAPHY

I received my B.S. in the Unified Honors Program sponsored by the Ford Foundation and my M.S. in math from Polytechnic University. After teaching my first class, as a substitute for a full professor, I heard one student say to another upon leaving the classroom, "At least we have someone who can teach the stuff." I was hooked forever on teaching. Since then, I have taught at C.U.N.Y., Westfield State College, Rutgers, and Poly. No matter how I feel, I always feel a lot better when I teach. I always feel great when students tell me they used to hate math or couldn't do math and now they like it more and can do it better.

My main blessing is my family. I have a fantastic wife in Marlene. My kids are wonderful: daughter Sheryl, son Eric, son-in-law Glenn, and daughter-in-law Wanda. My grandchildren are terrific: Kira, Evan, Sean, Sarah, and Ethan. My hobbies are golf, bowling, bridge, crossword puzzles, and Sudoku. My ultimate goals are to write a book to help parents teach their kids math, a high school text that will advance our kids' math abilities, and a calculus text students can actually understand.

To me, teaching is always a great joy. I hope that I can give some of that joy to you. I do know this book will help you get the score you need to get into the graduate school of your choice.

OTHER BOOKS

Bob Miller's Math for the GRE

Bob Miller's Basic Math and Pre-Algebra for the Clueless, Second Edition

Bob Miller's Algebra for the Clueless, Second Edition

Bob Miller's Geometry for the Clueless, Second Edition

Bob Miller's Math SAT for the Clueless, Second Edition

Bob Miller's Pre-Calc with Trig for the Clueless, Third Edition

Bob Miller's High School Calc for the Clueless

Bob Miller's Calc 1 for the Clueless, Second Edition

Bob Miller's Calc 2 for the Clueless, Second Edition

Bob Miller's Calc 3 for the Clueless

ABOUT RESEARCH & EDUCATION ASSOCIATION

Founded in 1959, Research & Education Association (REA) is dedicated to publishing the finest and most effective educational materials—including software, study guides, and test preps—for students in middle school, high school, college, graduate school, and beyond.

REA's test preparation series includes books and software for all academic levels in almost all disciplines. REA publishes test preps for students who have not yet entered high school, as well as high school students preparing to enter college. Students from countries around the world seeking to attend college in the United States will find the assistance they need in REA's publications. For college students seeking advanced degrees, REA publishes test preps for many major graduate school admission examinations in a wide variety of disciplines, including engineering, law, and medicine. Students at every level, in every field, with every ambition can find what they are looking for among REA's publications.

REA's publications and educational materials are highly regarded and continually receive an unprecedented amount of praise from professionals, instructors, librarians, parents, and students. Our authors are as diverse as the subject matter represented in the books we publish. They are well known in their respective disciplines and serve on the faculties of prestigious colleges and universities throughout the United States and Canada.

Today, REA's wide-ranging catalog is a leading resource for teachers, students, and professionals.

We invite you to visit us at *www.rea.com* to find out how "REA is making the world smarter."

REA ACKNOWLEDGMENTS

In addition to our author, we would like to thank Larry B. Kling, Vice President, Editorial, for his overall direction; Pam Weston, Vice President, Publishing, for setting the quality standards for production integrity and managing the publication to completion; Michael Reynolds, Managing Editor, for editorial contributions and project management; Mel Friedman, Lead Mathematics Editor, for proofreading; Rachel DiMatteo, Graphic Designer, for designing this book; Christine Saul, Senior Graphic Artist, for designing our cover; Jeff LoBalbo, Senior Graphic Designer, for coordinating pre-press electronic file mapping, and Aquent Publishing Services, for typesetting this edition. Photo by Eric L. Miller.

ABOUT THIS BOOK

Congratulations to you who have graduated or are about to graduate college! You are about to begin another great adventure. Before the journey starts, you must take the GMAT. This book is designed for you to maximize your score on the quantitative section, the math. This book teaches you the skills you need for the GMAT, some of which you may have forgotten. It then gives numerous problems that are typical of this test. I wrote this book in a way so that you will enjoy it and to help relieve some of your anxiety about the test.

The propaganda about the GMAT says that the test attempts to find out your knowledge of business, your job and interpersonal skills at the beginning of your undergraduate work, and subjective skills such as motivation and creativity. This, of course, refers to the English sections as well as the math sections. The math skills that are required are no more than those learned in Algebra II.

An interesting twist is that you cannot use scratch paper to work out answers. The following items are not permitted: notes, scratch paper (again), calculators or watch calculators, stop watches or watch alarms, personal data assistants (PDAs), telephones or cells, beepers or pages, photographic devices, stereos, radios or TVs, any other electronic aid that could help you, books or pamphlets, dictionaries, translators or thesauri, pens or any other writing devices, rulers or any other measuring devices. It other words, you are taking the test on computer by yourself.

Because of these conditions, you must memorize the formulas that you will use for this test. In addition, you must do all the arithmetic in your head. This isn't quite as scary as it sounds. Because you have to do everything in your head, the algebraic manipulations are not too complicated. Complicated arithmetic is not on this test. This book will show you ways to minimize how much arithmetic you need or, in some cases, eliminate it completely. However the test will require that you understand the material. You do have to be creative in coming up with some of the solutions—thinking "outside the box." This test really appeals to me because I love puzzles.

There are two kinds of questions on the GMAT. The first is the same kind of questions you took on the SAT. There is a problem to solve with five answer choices. The second kind of question is new: data sufficiency. Chapter 15 discusses this kind of question in detail (you can peek if you want to see more now). Briefly, the problems ask whether there is sufficient information given to solve a problem. In virtually all of the examples, you solve nothing; but you do have to know the facts of the question. This is a very important skill in the real world. It is tremendously important to know when you have enough information to solve a problem or whether you need more.

The GMAT is now a computer-adaptive test (CAT), given in English. The math section gives 75 minutes for 37 questions, approximately two minutes a question. You are given a question of moderate difficulty. When you know the answer, enter it. If your answer is correct, you will be given a harder question. Otherwise, you will be given an easier one. BE CAREFUL TO CHOOSE THE ANSWER YOU THINK IS CORRECT. It would be awful to get a lower score than you deserve only because you hit the wrong key! You must pace yourself, since failure to finish the 37 questions will result in a significantly lower score. If there is a question you absolutely don't know, you must guess. If wrong, the next question will be easier. If you answer it correctly, you will return to a similar level of difficulty. Getting those correct will result it harder questions. By the end of the test, the computer closes in on a score that best approximates your ability. You will probably take the GMAT in a room with many other people. However, the GMAT has a vast resource of questions. The people to your left and right will be taking different questions to arrive at their scores.

You must focus on your goal: it is to get a score that will get you into the MBA program of your choice. However, please keep in mind the following: When you get your MBA, the degree itself only gets your first job and first salary. If you go to a more prestigious school, your beginning salary of your first job will be higher, and so will be their expectations of you. Once you get your first job, what counts is your ability to do your job well!

As I said, I really like GMAT questions. In fact, I like all kinds of puzzles, both mathematical and word. To me the GMAT is a game. When you win, you win the graduate school of your choice.

Good luck!

CHAPTER 1: *The Basics*

" *All math begins with whole numbers.
Master them and you will begin to speak
the language of math.* "

Let's begin at the beginning. The GMAT works only with real numbers, numbers that can be written as decimals. However, it does not always say "numbers." Let's get specific.

NUMBERS

Whole numbers: 0, 1, 2, 3, 4, …

Integers: 0, ±1, ±2, ±3, ±4, …, where ±3 stands for both $+3$ and -3.

Positive integers are integers that are greater than 0. In symbols, $x > 0$, x an integer

Negative integers are integers that are less than 0. In symbols, $x < 0$, x an integer

Even integers: 0, ±2, ±4, ±6, …

Odd integers: ±1, ±3, ±5, ±7, …

Inequalities

For any numbers represented by a, b, c, or d on the number line:

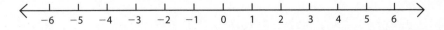

We say $c > d$ (c is greater than d) if c is to the right of d on the number line.

We say $d < c$ (d is less than c) if d is to the left of c on the number line.

$c > d$ is equivalent to $d < c$

$a \leq b$ means $a < b$ or $a = b$; likewise, $a \geq b$ means $a > b$ or $a = b$.

Example 1: $4 \leq 7$ is true, since $4 < 7$; $9 \leq 9$ is true, since $9 = 9$; but $7 \leq 2$ is false, since $7 > 2$.

Example 2:   Find all integers between −4 and 5.

Solution: {−3, −2, −1, 0, 1, 2, 3, 4}.

Notice that the word "between" does *not* include the endpoints.

Example 3:   Graph all the multiples of five between 20 and 40 inclusive.

Solution:

Notice that "inclusive" means to include the endpoints.

Odd and Even Numbers

Here are some facts about odd and even integers that you should know.

- The sum of two even integers is even.
- The sum of two odd integers is even.
- The sum of an even integer and an odd integer is odd.
- The product of two even integers is even.
- The product of two odd integers is odd.
- The product of an even integer and an odd integer is even.
- If n is even, n^2 is even. If n^2 is even and n is an integer, then n is even.
- If n is odd, n^2 is odd. If n^2 is odd and n is an integer, then n is odd.

OPERATIONS ON NUMBERS

Product is the answer in multiplication; **quotient** is the answer in division; **sum** is the answer in addition; and **difference** is the answer in subtraction.

Since $3 \times 4 = 12$, 3 and 4 are said to be **factors** or **divisors** of 12, and 12 is both a **multiple** of 3 and a **multiple** of 4.

A prime is a positive integer with exactly two distinct factors, itself and 1. The number 1 is not a prime since only $1 \times 1 = 1$. It might be a good idea to memorize the first eight primes:

2, 3, 5, 7, 11, 13, 17, and 19.

The number 4 has more than two factors: 1, 2, and 4. Numbers with more than two factors are called **composites**. The number 28 is a **perfect** number since if we add the factors less than 28, they add to 28.

Example 4:   Write all the factors of 28.

Solution: 1, 2, 4, 7, 14, and 28.

Example 5:   Write 28 as the product of prime factors.

Solution: $28 = 2 \times 2 \times 7$.

Example 6:   Find all the primes between 70 and 80.

Solution: 71, 73, 79. How do we find this easily? First, since 2 is the only even prime, we only have to check the odd numbers. Next, we have to know the divisibility rules:

- A number is divisible by 2 if it ends in an even number. We don't need this here because then it can't be prime.

- A number is divisible by 3 (or 9) if the sum of the digits is divisible by 3 (or 9). For example, 456 is divisible by 3 since the sum of the digits is 15, which is divisible by 3 (it's not divisible by 9, but that's okay).

- A number is divisible by 4 if the number named by the last two digits is divisible by 4. For example, 3936 is divisible by 4 since 36 is divisible by 4.

- A number is divisible by 5 if the last digit is 0 or 5.

- The rule for 6 is a combination of the rules for 2 and 3.

- It is easier to divide by 7 than to learn the rule for 7.

- A number is divisible by 8 if the number named by the last *three* digits are divisible by 8.

- A number is divisible by 10 if it ends in a zero, as you know.

- A number is divisible by 11 if the difference between the sum of the even-place digits (2nd, 4th, 6th, etc.) and the sum of the odd-place digits (1st, 3rd, 5th, etc.) is a multiple of 11. For example, for the number 928,193,926: the sum of the odd digits (9, 8, 9, 9, and 6) is 41; the sum of the even digits (2, 1, 3, 2) is 8; and $41 - 8$ is 33, which is divisible by 11. So 928,193,926 is divisible by 11.

That was a long digression!!!!! Let's get back to example 6.

We have to check only 71, 73, 75, 77, and 79. The number 75 is not a prime since it is ends in a 5. The number 77 is not a prime since it is divisible by 7. To see if the other three are prime, for any number less than 100 you have to divide by the primes 2, 3, 5, and 7 only. You will quickly find they are primes.

Rules for Operations on Numbers

Note *( ) are called parentheses (singular: parenthesis); [ ] are called brackets; { } are called braces.*

Rules for adding signed numbers

1. If all the signs are the same, add the numbers and use that sign.

2. If two signs are different, subtract them, and use the sign of the larger numeral.

Example 7: **a.** $3 + 7 + 2 + 4 = +16$ **c.** $5 - 9 + 11 - 14 = 16 - 23 = -7$

b. $-3 - 5 - 7 - 9 = -24$ **d.** $2 - 6 + 11 - 1 = 13 - 7 = +6$

Rules for multiplying and dividing signed numbers
Look at the minus signs only.

1. Odd number of minus signs—the answer is minus.

2. Even number of minus signs—the answer is plus.

Example 8: $\dfrac{(-4)(-2)(-6)}{(-2)(+3)(-1)} =$

Solution: Five minus signs, so the answer is minus, -8.

Rule for subtracting signed numbers
The sign $(-)$ means subtract. Change the problem to an addition problem.

Example 9: **a.** $(-6) - (-4) = (-6) + (+4) = -2$

b. $(-6) - (+2) = (-6) + (-2) = -8$, since it is now an adding problem.

Order of Operations

In doing a problem such as $4 + 5 \times 6$, the **order of operations** tells us whether to multiply or add first:

1. If given letters, substitute in parentheses the value of each letter.

2. Do operations in parentheses, inside ones first, and then the tops and bottoms of fractions.

3. Do exponents next. (Chapter 3 discusses exponents in more detail.)

4. Do multiplications and divisions, left to right as they occur.

5. The last step is adding and subtracting. Left to right is usually the safest way.

Example 10: $4 + 5 \times 6 =$

Solution: $4 + 30 = 34$

Example 11: $(4 + 5)6 =$

Solution: $(9)(6) = 54$

Example 12: $1000 \div 2 \times 4 =$

Solution: $(500)(4) = 2{,}000$

Example 13: $1000 \div (2 \times 4) =$

Solution: $1000 \div 8 = 125$

Example 14: $4[3 + 2(5 - 1)] =$

Solution: $4[3 + 2(4)] = 4[3 + 8] = 4(11) = 44$

Example 15: $\dfrac{3^4 - 1^{10}}{4 - 10 \times 2} =$

Solution: $\dfrac{81 - 1}{4 - 20} = \dfrac{80}{-16} = -5$

Example 16: If $x = -3$ and $y = -4$, find the value of:

    a. $7 - 5x - x^2$

    b. $xy^2 - (xy)^2$

Solutions: **a.** $7 - 5x - x^2 = 7 - 5(-3) - (-3)^2 = 7 + 15 - 9 = 13$

    b. $xy^2 - (xy)^2 = (-3)(-4)^2 - ((-3)(-4))^2 = (-3)(16) - (12)^2$

             $= -48 - 144 = -192$

Before we get to the exercises, let's talk about ways to describe a group of numbers (data).

DESCRIBING DATA

Four of the measures that describe data are used on the GMAT. The first three are measures of central tendency; the fourth, the range, measures the span of the data. Chapter 14 discusses these measures in more detail.

Mean: Also called average. Add up the numbers and divide by how many numbers you have added up.

Median: Middle number. Put the numbers in numeric order and see which one is in the middle. If there are two "middle" numbers take the average of them. This happens with an even number of data points.

Mode: Most common. Which number(s) appears the most times? A set with two modes is called bimodal. There can actually be any number of modes, including that everything is a mode.

Range: Highest number minus the lowest number.

Example 17: Find the mean, median, mode, and range for 5, 6, 9, 11, 12, 12, and 14.

Solutions: Mean: $\dfrac{5 + 6 + 9 + 11 + 12 + 12 + 14}{7} = \dfrac{69}{7} = 9\dfrac{6}{7}$

Median: 11

Mode: 12

Range: $14 - 5 = 9$

Example 18: Find the mean, median, mode, and range for 4, 4, 7, 10, 20, 20.

Solutions: Mean: $\dfrac{4 + 4 + 7 + 10 + 20 + 20}{6} = \dfrac{65}{6} = 10\dfrac{5}{6}$

Median: For an even number of points, it is the mean of the middle two:

$\dfrac{7 + 10}{2} = 8.5$

Mode: There are two: 4 and 20 (blackbirds?)

Range: $20 - 4 = 16$

Example 19: Jim received 83 and 92 on two tests. What grade must the third test be in order to have an average (mean) of 90?

Solution: There are two methods.

Method 1: To get a 90 average on three tests Jim needs $3(90) = 270$ points. So far, he has $83 + 92 = 175$ points. So, Jim needs $270 - 175 = 95$ points on the third test.

Method 2 (my favorite): 83 is -7 from 90. 92 is $+2$ from 90. $-7 + 2 = -5$ from the desired 90 average. Jim needs $90 + 5 = 95$ points on the third test. (Jim needs to "make up" the 5-point deficit, so add it to the average of 90.)

It is essential to learn how to do one of these methods well because calculators are not permitted for the GMAT. For this exam, the second method is really better, but it is always your choice which method you can do easier and faster.

(Q) **Finally, after a long introduction, we get to some exercises.**

The GMAT has two types of questions. Until Chapter 15, we will do only the standard multiple-choice questions, the ones you figure out and choose the correct answer.

Exercise 1: If $x = -5$, the value of $-3 - 4x - x^2$ is $-3 + 20 - 25$

  A. -48 D. 13

  B. -8 E. 4

  C. 2

Exercise 2: $-0(2) - \dfrac{0}{2} - 2 =$

  A. 0   D. -6

  B. -2 E. Undefined

  C. -4

(handwritten at top) $380 - 279 = x = 101$

(handwritten) $95 = \dfrac{(90 + 91 + 98 + x)}{4}$

Exercise 3: The scores on three tests were 90, 91, and 98. What does the score on the fourth test have to be in order to get exactly a 95 average (mean)?

    A. 97   D. 100

    B. 98   (E.) Not possible

    C. 99

Exercise 4: On a true-false test, 20 students scored 90, and 30 students scored 100. The sum of the mean, median, and mode is

    A. 300   D. 294

    (B.) 296   E. 275

    C. 295

(handwritten margin) 90 90 : : 100 100 ...

(handwritten) 1800 / 3000 ; 4800 / 50 stud. total ; median = 100 ; mean = 96 ; mode = 100

Exercise 5: On a test, m students received a grade of x, n students received a grade of y, and p students received a grade of z. The average (mean) grade is:

    A. $\dfrac{mxnypz}{xyz}$ (D.) $\dfrac{mx + ny + pz}{m + n + p}$

    B. $\dfrac{mx + ny + pz}{x + y + z}$ E. $mnp + \dfrac{xyz}{x + y + z}$

    C. $\dfrac{mx + ny + pz}{xyz}$

(handwritten) $\dfrac{mx + ny + pz}{m + n + p}$

Exercise 6: The largest positive integer in the following list that divides evenly into 2,000,000,000,000,003 is

    A. 33   D. 3

    B. 11   (E.) 1

    C. 10

Exercise 7: If m and n are odd integers, which of the following is odd?

(handwritten) $4 \times 3 = 12$

    A. $mn + 3$ D. $(m + 1)(n - 2)$

    B. $m^2 + (n + 2)^2$ E. $m^4 + m^3 + m^2 + m$

    (C.) $mn + m + n$

Exercise 8: If $(m + 3)$ is a multiple of 4, which of these is also a multiple of 4?

    A. $m - 3$ D. $m + 9$

    B. m (E.) $m + 11$

    C. $m + 4$

(handwritten margin) 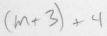 $(m + 3) + 4$

Exercise 9: If p and q are primes, which one of the following *can't* be a prime?

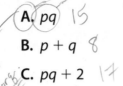

    A. pq

    B. $p + q$

    C. $pq + 2$

    D. $2pq + 1$

    E. $p^2 + q^2$

Exercise 10: The sum of the first n positive integers is p. In terms of n and p, what is the sum of the next n positive integers?

    A. np

    B. $n + p$

    C. $n^2 + p$

    D. $n + p^2$

    E. $2n + 2p$

Exercise 11: Let p be prime, with $20p$ divisible by 6; p could be

    A. 3

    B. 4

    C. 5

    D. 6

    E. 7

 Let's look at the answers.

Answer 1: B: $-3 - 4(-5) - (-5)^2 = -3 + 20 - 25 = -8$.

Answer 2: B: $0 - 0 - 2 = -2$.

Answer 3: E: $95(4) = 380$ points; $90 + 91 + 98 = 279$ points. The fourth test would have to be $380 - 279 = 101$.

Answer 4: B: The median is 100; the mode is 100; for the mean we can use 2 and 3 instead of 20 and 30 since the ratio is the same: $\dfrac{2(90) + 3(100)}{5} = 96$.

Answer 5: D: $Mean = \dfrac{\text{total value of items}}{\text{total number of items}} = \dfrac{mx + ny + pz}{m + n + p}$.

Answer 6: E: The number is not divisible by 11 because $3 - 2 = 1$, which is not a multiple of 11. It is not divisible by 3 because $3 + 2 = 5$ is not divisible by 3. Since is it not divisible by 11 or 3, it is not divisible by 33. Finally, it is not divisible by 10 because it does not end in a 0. So the answer is E because all numbers are divisible by 1.

Answer 7:   **C:** Only C is the sum of three odd integers. All of the other answer choices are even.

Answer 8:   **E:** If $m + 3$ is a multiple of 4, then $m + 3 + 8$ is a multiple of 4 since 8 is a multiple of 4.

Answer 9:   **A:** By substituting proper primes, all the others might be prime.

Answer 10:   **C:** Say $n = 5$; $p = 1 + 2 + 3 + 4 + 5$. The next five are $(1 + 5) + (2 + 5) + (3 + 5) + (4 + 5) + (5 + 5) = p + n^2$.

Answer 11:   **A:** 3 and 6 will work, but only 3 is a prime.

CHAPTER 2: *Decimals, Fractions, and Percentages*

"One must master the parts as well as the whole to fully understand."

Since the GMAT does not allow calculators, you might be spending a little more time on this chapter than you planned. Let's start with decimals.

DECIMALS

Rule 1: When adding or subtracting, line up the decimal points.

Example 1: Add: 3.14 + 234.7 + 86

Solution:
$$\begin{array}{r} 3.14 \\ 234.7 \\ +\ 86. \\ \hline 323.84 \end{array}$$

Example 2: Subtract: 56.7 − 8.82

Solution:
$$\begin{array}{r} 56.70 \\ -\ 8.82 \\ \hline 47.88 \end{array}$$

Rule 2: In multiplying numbers, count the number of decimal places and add them. In the product, this will be the number of decimal places for the decimal point.

Example 3: Multiply 45.67 by .987.

Solution: The answer will be 45.07629. You will need to know the answer has five decimal places.

Example 4: Multiply 2.8 by .6:

Solution: The answer is 1.68. The GMAT will expect you to do this problem. Let's try another.

Example 5: What is the value of $2b^2 - 2.4b - 1.7$ if $b = .7$?

Solution: The standard way to do this type of problem is to directly substitute.
$2(.7)(.7) = 2(.49) = .98; (-2.4)(.7) = -1.68; .98 - 1.68 - 1.7 = -2.40.$

However, you do not have a paper and pencil on this test, and you will be given answers from which to choose, so an approximation works just as well and is quicker. $(.7)(.7) = .49$; times 2 is .98, or approximately 1. -2.4 times $.7 = 1.68$ or approximately 1.7; $1 - 1.7 = -.7$; added to -1.7 is -2.4. In this case, we get the same numerical value. If we didn't, our approximate answer would still probably be closer to the correct answer than to the other choices.

In business and on this test, one skill you must have to save time is the ability to make reasonable approximations. This book will point out problems that can be approximated.

Rule 3: When you divide, move the decimal point in the divisor and the dividend the same number of places.

Example 6: Divide 23.1 by .004.

Solution: In our heads, we write the problem as $\dfrac{23.1}{.004}$. Multiplying the numerator and denominator by 1000, we get $\dfrac{23.1}{.004} = \dfrac{23.1 \times 1000}{.004 \times 1000} = \dfrac{23100}{4} = $ 5775. When we multiply by 1, the fraction doesn't change. Since $\dfrac{1000}{1000} = 1$, the fraction is the same.

Rule 4: When reading a number with a decimal, read the whole part, only say the word "and" when you reach the decimal point, then read the part after the decimal point as if it were a whole number, and say the last decimal place. Whew!

Example 7:

Number	Read
4.3	Four and three tenths
2,006.73	Two-thousand six and seventy-three hundredths
1,000,017.009	One million seventeen and nine thousandths

Q **Let's do some exercises.**

Exercise 1: Which is smallest:

A. .04   **D.** .04444

B. .0401   **E.** .041

C. .04001

Exercise 2: $\dfrac{70000}{100000} + \dfrac{800}{10,00000} + \dfrac{9}{1,000,000} = \dfrac{70,809}{1,000,000}$

A. 789   **D.** .070809

B. .0789   **E.** 078009

C. .07089

A **Let's look at the answers.**

Answer 1: A: All have the same tenths (0) and hundredths (4). The rest of the places of A. are zero.

Answer 2: D: Hundredths is the second decimal place, ten-thousandths in the fourth, and millionths is the sixth place.

Now, let's go over fractions.

FRACTIONS

The top of a fraction is called the **numerator**; the bottom is the **denominator**.

Rule 1: If the bottoms of a fraction are the same, the bigger the top, the bigger the fraction.

Example 8: Suppose I am a smart first grader. Can you explain to me which is bigger, $\dfrac{3}{5}$ or $\dfrac{4}{5}$?

Solution: Suppose we have a pizza pie. Then $\dfrac{3}{5}$ means we divide a pie into 5 equal parts, and I get 3. And $\dfrac{4}{5}$ means I get 4 pieces out of 5. So $\dfrac{3}{5} < \dfrac{4}{5}$.

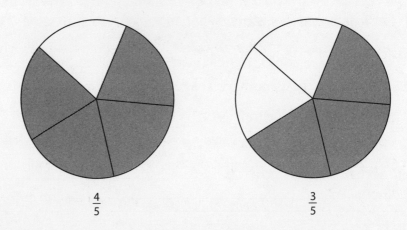

$$\frac{4}{5} \qquad\qquad \frac{3}{5}$$

Rule 2: If the tops of two fractions are the same, the bigger the bottom, the smaller the fraction.

Example 9:   Which fraction, $\frac{3}{5}$ or $\frac{3}{4}$, is bigger?

Solution: Use another pizza pie example. In comparing $\frac{3}{5}$ and $\frac{3}{4}$, we get the same number of pieces (3). However, if the pie is divided into 4, the pieces are bigger, so $\frac{3}{5} < \frac{3}{4}$.

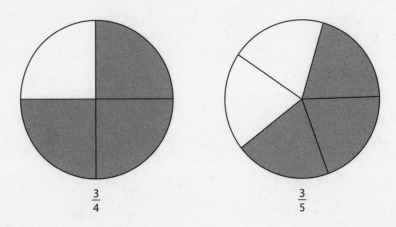

$$\frac{3}{4} \qquad\qquad \frac{3}{5}$$

Rule 3: If the tops and bottoms are different, find the Least Common Denominator (LCD) and compare the tops.

Before we get into this section, the teacher in me (and maybe the purist in you) must tell you we really are talking about rational numbers, not fractions. There are two definitions.

Definition 1: A **rational** number is any integer divided by an integer, with the denominator not equaling zero.

Definition 2: A **rational** number is any repeating or terminating decimal.

Technically, $\frac{\pi}{6}$ *is a fraction but not a rational number. We will use the term "fraction" here instead of "rational number." If it is negative, we will say "negative fraction."*

Note the following facts about fractions:

- $3 < 4$, but $-3 > -4$. Similarly, $\frac{3}{5} < \frac{4}{5}$, but $-\frac{3}{5} > -\frac{4}{5}$. We will do more of this later.

- A fraction (positive) is bigger than 1 if the numerator is bigger than the denominator.

- A fraction is less than $\frac{1}{2}$ if the denominator is more than twice the numerator.

- To double a fraction, either multiply the numerator by two or divide the denominator by 2.

- Adding the same positive number to the numerator and denominator makes the fraction closer to one.

- Two fractions are **equivalent** if they can be reduced to the same fraction. $\frac{6}{9}$ and $\frac{10}{15}$ are equivalent because they both reduce to $\frac{2}{3}$.

Ⓠ   **Let's do some more exercises.**

Exercise 3:   Which fraction is largest?

A. $\frac{7}{4}$ $\div 25$ 175
     $\times 25 = \frac{175}{100}$

B. $\frac{11}{8}$ $\times \frac{125}{12.5} = \frac{1375}{100}$

C. $\frac{3}{2}$ $\times \frac{50}{50} = \frac{150}{100}$

D. $\frac{12}{5}$ $\times \frac{20}{20} = \frac{240}{100}$

E. $\frac{3}{100}$

Exercise 4:   Which fraction is smallest?

A. $\frac{3}{5}$ 0.6

B. $\frac{5}{11}$ $.45$

C. $\frac{7}{13}$ $.538$

D. $\frac{9}{16}$ 0.5625

E. $\frac{100}{199}$ $.5025$

Exercise 5:   Which fraction is largest?

A. $\dfrac{1}{3}$ ≈ 0.33

D. $\dfrac{3}{.1}$ ≈

B. $\dfrac{1}{.3}$ ≈ 3

E. $\dfrac{3}{(.1)^2}$ ≈ 3

C. $\dfrac{1}{(.3)^2}$ ≈ 9

Exercise 6:   If $100 \leq x \leq 10{,}000$ and $.0001 \leq y \leq .01$, the smallest possible value of $\dfrac{x}{y}$ is

A. 10,000

D. 10,000,000

B. 100,000

E. 1,000,000,000

C. 1,000,000

Exercise 7:   Suppose we have $\dfrac{x+y}{x-y}$, where $8 \leq x \leq 10$ and $2 \leq y \leq 4$. The maximum possible value for this fraction is

A. $\dfrac{3}{2}$

D. 3

⑧ $\dfrac{12}{4}$ = 3

B. $\dfrac{5}{3}$

E. 6

⑨ $\dfrac{13}{5}$ = 2.6

C. $\dfrac{7}{3}$

⑩ $\dfrac{14}{6}$ = 2.3̄

 Let's look at the answers.

Answer 3:   D: It is the only fraction for which the top is more than double the bottom, so it is the only fraction with a value greater than 2.

Answer 4:   B: It is the only fraction less than one-half; the bottom is more than double the top.

Answer 5:   E: Answer choice A is less than 1. Comparing the other choices, C is bigger than B, and E is bigger than D. Since E has a bigger top and smaller bottom than C, it is larger. We will explore this in greater detail in the next chapter.

Answer 6:   A: To make a fraction as small as possible, we need a fraction with the smallest top ($x = 100$) and the biggest bottom ($y = .01$), or $\dfrac{100}{.01} = 10{,}000.$

Answer 7:   D: This one is not so simple. In this case, the extreme values, min and max occur at the "ends." We must try $x = 8$ and 10, $y = 2$ and 4, and all combinations of these numbers. In this case, the max occurs when $x = 8$ and $y = 4$.

Adding and Subtracting Fractions

If the denominators are the same, add or subtract the tops, keep the bottom the same, and reduce if necessary.

$$\frac{7}{43} + \frac{11}{43} - \frac{2}{43} = \frac{16}{43}$$

$$\frac{2}{9} + \frac{4}{9} = \frac{6}{9} = \frac{2}{3}$$

$$\frac{a}{m} + \frac{b}{m} - \frac{c}{m} = \frac{a + b - c}{m}$$

There is much more to talk about if the denominators are unlike.

The quickest way to add (or subtract) fractions with different denominators, especially if they contain letters or the denominators are small (but different), is to multiply the top and bottom of each fraction by the least common multiple, LCM. This consists of three words: multiple, common, and least.

Example 10:   What is the LCM of 6 and 8?

Solution: **Multiples** of 6 are 6, 12, 18, 24, 30, 36, 42, 48, 54, 60, 66, 72, 78,…

Multiples of 8 are 8, 16, 24, 32, 40, 48, 56, 64, 72, 80,…

Common multiples of 6 and 8 are 24, 48, 72, 96, 120,…

The **least common multiple** of 6 and 8 is 24.

When adding or subtracting fractions, multiply the top and bottom of each fraction by the LCM divided by the denominator of that fraction:

$$\frac{a}{b} - \frac{x}{y} = \left(\frac{a}{b} \times \frac{y}{y}\right) - \left(\frac{x}{y} \times \frac{b}{b}\right) = \frac{ay}{by} - \frac{bx}{by} = \frac{ay - bx}{by}$$

$$\frac{7}{20} - \frac{3}{11} = \left(\frac{7}{20} \times \frac{11}{11}\right) - \left(\frac{3}{11} \times \frac{20}{20}\right) = \frac{7(11) - 3(20)}{20(11)} = \frac{17}{220}$$

On the GMAT, you must be able to perform these calculations quickly.

You also must be able to add in your head problems such as $\frac{3}{8} + \frac{5}{6} = \frac{9}{24} + \frac{20}{24} = \frac{29}{24}$.

Also, you need to be able to do the following in your head: $\frac{3}{4} + \frac{5}{6} = \frac{9}{12} + \frac{10}{12} = \frac{19}{12}$.

Multiplication of Fractions

To multiply fractions, multiply the numerators and multiply the denominators, reducing as you go. With multiplication, it is *not* necessary to have the same denominators.

$$\frac{3}{7} \times \frac{4}{11} = \frac{12}{77}$$

$$\frac{a}{b} \times \frac{c}{d} = \frac{a \times c}{b \times d}$$

Division of Fractions

To divide fractions, invert the second fraction and multiply, reducing if necessary. To **invert** a fraction means to turn it upside down. The new fraction is called the **reciprocal** of the original fraction. So the reciprocal of $\frac{2}{3}$ is $\frac{3}{2}$; the reciprocal of -5 is $-\frac{1}{5}$; and the reciprocal of a is $\frac{1}{a}$ if $a \neq 0$.

For example,

$$\frac{3}{4} \div \frac{11}{5} = \frac{3}{4} \times \frac{5}{11} = \frac{15}{44}$$

$$\frac{m}{n} \div \frac{p}{q} = \frac{m}{n} \times \frac{q}{p} = \frac{m \times q}{n \times p}$$

$$\frac{1}{4} \div 5 = \frac{1}{4} \times \frac{1}{5} = \frac{1}{20}$$

Let's do a few problems. Do them without paper and pencil.

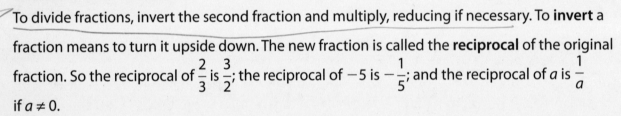

Example 11: <u>Problem</u>   <u>Solution</u>

a. $\frac{7}{9} - \frac{3}{22} =$ $\frac{127}{198}$

b. $\frac{3}{4} + \frac{5}{6} - \frac{1}{8} =$ $\frac{35}{24}$, or $1\frac{11}{24}$

c. $\dfrac{3}{10} + \dfrac{2}{15} - \dfrac{4}{5} =$ $\dfrac{-11}{30}$

d. $\dfrac{1}{4} + \dfrac{1}{8} + \dfrac{7}{16} =$ $\dfrac{13}{16}$

e. $2 + \dfrac{2}{3} + \dfrac{2}{9} + \dfrac{2}{27} =$ $2\dfrac{26}{27}$, or $\dfrac{80}{27}$

f. $\dfrac{5}{24} - \dfrac{7}{18} =$ $-\dfrac{13}{72}$

g. $\dfrac{10}{99} - \dfrac{9}{100} =$ $\dfrac{109}{9900}$

h. $\dfrac{7}{9} \times \dfrac{5}{3} =$ $\dfrac{35}{27}$, or $1\dfrac{8}{27}$

i. $\dfrac{11}{12} \div \dfrac{9}{11} =$ $\dfrac{121}{108}$, or $1\dfrac{13}{108}$

j. $\dfrac{5}{9} \times \dfrac{6}{7} =$ $\dfrac{10}{21}$

k. $\dfrac{12}{13} \div \dfrac{8}{39} =$ $\dfrac{9}{2}$, or $4\dfrac{1}{2}$

l. $\dfrac{10}{12} \div \dfrac{15}{40} =$ $\dfrac{20}{9}$, or $2\dfrac{2}{9}$

m. $\dfrac{2}{3} \div 12 =$ $\dfrac{1}{18}$

n. $\dfrac{2}{3} \times \dfrac{3}{4} \times \dfrac{4}{5} \times \dfrac{5}{6} \times \dfrac{6}{7} =$ $\dfrac{2}{7}$

o. $\dfrac{5}{8} \times \dfrac{7}{6} \div \dfrac{35}{24} =$ $\dfrac{1}{2}$

Q **Let's do some more exercises.**

Exercise 8: What number when multiplied by $\frac{3}{4}$ gives $\frac{7}{8}$?

  A. $\frac{21}{32}$ D. $\frac{8}{7}$

  B. $\frac{6}{5}$ E. $\frac{4}{3}$

  C. $\frac{7}{6}$

Exercise 9: The average (mean) of $\frac{1}{4}$ and $\frac{1}{8}$ is

  A. $\frac{1}{12}$ D. $\frac{5}{32}$

  B. $\frac{1}{6}$ E. $\frac{7}{24}$

  C. $\frac{3}{16}$

A **Let's look at the answers.**

Answer 8: **C:** We need to solve $\frac{3}{4}x = \frac{7}{8}$. Then $x = \frac{7}{8} \times \frac{4}{3} = \frac{7}{6}$. I hate to keep emphasizing this, but you need to be able to do this is your head. The GMAT seems to like this kind of question.

Answer 9: **C:** $\frac{\left(\frac{1}{4} + \frac{1}{8}\right)}{2} = \frac{\left(\frac{3}{8}\right)}{2} = \frac{3}{16}.$

Changing from Decimals to Fractions and Back

To change from a decimal to a fraction, we read it and write it.

Example 12: Change 4.37 to a fraction.

Solution: We read it as 4 and 37 hundredths: $4\frac{37}{100} = \frac{437}{100}$, if necessary. That's it.

Example 13: Change to decimals:

a. $\dfrac{7}{4}$ b. $\dfrac{1}{6}$

Solution: For the fractions on the GMAT, the decimal will either terminate or repeat.

a. Divide 4 into 7.0000: $7.0000 \div 4 = 1.75$

b. Divide 6 into 1.0000: $1.0000 \div 6 = .1666\ldots = .1\overline{6}$

The bar over the 6 means it repeats forever; for example, $.3454545\ldots = .3\overline{45}$. means 45 repeats forever, but not the 3.

PERCENTAGES

% means hundredths: $1\% = \dfrac{1}{100} = .01$.

Follow these rules to change between percentages and decimals and fractions:

Rule 1: To change a percentage to a decimal, move the decimal point two places to the left and drop the % sign.

Rule 2: To change a decimal to a percentage, move the decimal point two places to the right and add a % sign.

Rule 3: To change from a percentage to a fraction, divide by 100% and simplify, or change the % sign to $\dfrac{1}{100}$ and multiply.

Rule 4: To change a fraction to a percentage, first change to a decimal, and then to a percentage.

Example 14: Change 12%, 4%, and .7% to decimals.

Solutions: $12\% = 12.\% = .12$; $4\% = 4.\% = .04$; $.7\% = .007$.

Example 15: Change .734, .2, and 34 to percentages.

Solutions: $.734 = 73.4\%$; $.2 = 20\%$; $34 = 34. = 3400\%$.

Example 16: Change 42% to a fraction.

Solution: $42\% = \dfrac{42}{100} = \dfrac{21}{50}$, or $42\% = 42 \times \dfrac{1}{100} = \dfrac{42}{100} = \dfrac{21}{50}$

Example 17: Change $\dfrac{7}{4}$ to a percentage.

Solution: $\dfrac{7}{4} = 1.75 = 175\%$

Two hundred years ago, when I was in elementary school, we had to learn the following decimal, fraction, and percent equivalents. Since there is no calculator allowed for the GMAT, it is essential to know the following:

Fraction	Decimal	Percentage	Fraction	Decimal	Percentage
$\dfrac{1}{2}$	.5	50%	$\dfrac{1}{8}$	.125	$12\dfrac{1}{2}\%$
$\dfrac{1}{4}$	.25	25%	$\dfrac{3}{8}$	.375	$37\dfrac{1}{2}\%$
$\dfrac{3}{4}$	.75	75%	$\dfrac{5}{8}$	.625	$62\dfrac{1}{2}\%$
$\dfrac{1}{5}$	.2	20%	$\dfrac{7}{8}$	.875	$87\dfrac{1}{2}\%$
$\dfrac{2}{5}$	.4	40%	$\dfrac{1}{16}$	.0625	$6\dfrac{1}{4}\%$
$\dfrac{3}{5}$	.6	60%	$\dfrac{3}{16}$	.1875	$18\dfrac{3}{4}\%$
$\dfrac{4}{5}$	.8	80%	$\dfrac{5}{16}$	.3125	$31\dfrac{1}{4}\%$
$\dfrac{1}{6}$	$.1\overline{6}$	$16\dfrac{2}{3}\%$	$\dfrac{7}{16}$	.4375	$43\dfrac{3}{4}\%$
$\dfrac{1}{3}$	$.\overline{3}$	$33\dfrac{1}{3}\%$	$\dfrac{9}{16}$	.5625	$56\dfrac{1}{4}\%$
$\dfrac{2}{3}$	$.\overline{6}$	$66\dfrac{2}{3}\%$	$\dfrac{11}{16}$	.6875	$68\dfrac{3}{4}\%$
$\dfrac{5}{6}$	$.8\overline{3}$	$83\dfrac{1}{3}\%$	$\dfrac{13}{16}$	.8125	$81\dfrac{1}{4}\%$
			$\dfrac{15}{16}$	.9375	$93\dfrac{3}{4}\%$

If you can't memorize all of these with denominator 16, memorize at least $\dfrac{1}{16}$ and $\dfrac{15}{16}$.

If you are good at doing percentage problems, skip this next section. Otherwise, here's a really easy way to do percentage problems. Make the following pyramid:

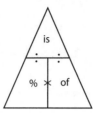

Example 18: What is 12% of 1.3?

Solution:

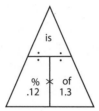

Put .12 in the % box (always change to a decimal in this box) and 1.3 in the "of" box. It tells us to multiply .12 × 1.3 = .156. That's all there is to it.

Example 19: 8% of what is 32?

Solution:

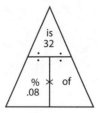

.08 goes in the % box. 32 goes in the "is" box. 32 ÷ .08 = 400.

Example 20: 9 is what % of 8?

Solution:

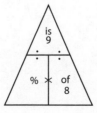

9 goes in the "is" box. 8 goes in the "of" box. 9 ÷ 8 × 100% = 112.5%.

The goal is to be able to do percentage problems without writing the pyramid.

Example 21:   In ten years, the population increases from 20,000 to 23,000. Find the actual increase and the percentage increase.

Solution: The actual increase is 23,000 − 20,000 = 3,000. The percentage increase is $\dfrac{3000}{20,000} \times 100\% = 15\%$ increase.

Example 22:   The cost of producing widgets decreased from 60 cents to 50 cents. Find the actual decrease and percentage decrease.

Solution: 60 − 50 = 10 cent decrease; $\dfrac{10}{60} = 16\dfrac{2}{3}\%$ decrease.

Note   *Percentage increases and decreases are figured on the original amount.*

Example 23:   The cost of a $2000 large-screen TV set is decreased by 30%. If there is 7% sales tax, how much does it cost?

Solution: $2000 × .30 = $600 discount. $2000 − $600 = $1400 cost. $1400 × .07 is $98. The total price is $1400 + $98 = $1498.

Note   *If we took 70% of $2000 (or 100% − 30%), we would immediately get the cost.*

There is an interesting story about why women wear miniskirts in London, England. It seems that the sales tax is $12\dfrac{1}{2}\%$ on clothes! But children's clothes are tax exempt. A girl's dress is any dress where the skirt is less than 24 inches, so that is why women in London wear miniskirts!

Q   **Let's do some exercises.**

Exercise 10:   The product of 2 and $\dfrac{1}{89}$ is:

A. $2\dfrac{1}{89}$ D. $\dfrac{2}{89}$

B. $1\dfrac{88}{89}$ E. $\dfrac{1}{172}$

C. 172

Exercise 11: $\dfrac{1}{50}$ of 2% of .02 is

A. .08

B. .008

C. .0008

D. .000008

E. 00000008

Exercise 12: 30% of 20% of a number is the same as 40% of what percentage of the same number?

A. 10

B. $12\dfrac{1}{2}$

C. 15

D. 18

E. Can't be determined without the number

Exercise 13: A fraction between $\dfrac{3}{43}$ and $\dfrac{4}{43}$ is

A. $\dfrac{1}{9}$

B. $\dfrac{3}{28}$

C. $\dfrac{5}{47}$

D. $\dfrac{7}{86}$

E. $\dfrac{9}{1849}$

Exercise 14: The reciprocal of $2 - \dfrac{3}{4}$ is

A. $\dfrac{1}{2} - \dfrac{4}{3}$

B. $-\dfrac{5}{4}$

C. $-\dfrac{4}{5}$

D. $\dfrac{5}{4}$

E. $\dfrac{4}{5}$

Exercise 15: A price increase of 20% followed by a decrease of 20% means the price is

A. Up 4%

B. Up 2%

C. The original price

D. Down 2%

E. Down 4%

Exercise 16: A price decreases 20% followed by a 20% increase. The final price is

A. Up 4%   D. Down 2%

B. Up 2%   E. Down 4%

C. The original price

Exercise 17: A 50% discount followed by another 50% discount gives a discount of

A. 25%   D. 90%

B. 50%   E. 100%

C. 75%

Exercise 18: $83\frac{1}{3}$% of $37\frac{1}{2}$% is

A. $\frac{1}{2}$ D. $\frac{5}{16}$

B. $\frac{2}{5}$ E. $\frac{3}{32}$

C. $\frac{1}{4}$

Exercise 19: Which is equivalent to .0625?

A. $\frac{3}{8}$ D. $\frac{1}{18}$

B. $\frac{1}{8}$ E. $\frac{1}{80}$

C. $\frac{1}{16}$

Exercise 20: If the fractions $\frac{19}{24}, \frac{1}{2}, \frac{3}{8}, \frac{3}{4}$, and $\frac{9}{16}$ were ordered least to greatest, the middle number would be

A. $\frac{19}{24}$ D. $\frac{3}{4}$

B. $\frac{1}{2}$ E. $\frac{9}{16}$

C. $\frac{3}{8}$

Exercise 21: $\dfrac{31}{125} =$

A. 0.320   D. 0.252

B. 0.310   E. 0.248

C. 0.288

 Let's look at the answers.

Answer 10:   D: $\dfrac{2}{1} \times \dfrac{1}{89} = \dfrac{2}{89}$.

Answer 11:   D: .02 × .02 × .02 = .000008.

Answer 12:   C: We can forget the number and forget the percentages. (30)(20) = (40)(?). ? = 15.

Answer 13:   D: $\dfrac{3}{43} = \dfrac{6}{86}$; $\dfrac{4}{43} = \dfrac{8}{86}$; between is $\dfrac{7}{86}$. As another example like this one, to get nine numbers between $\dfrac{3}{43}$ and $\dfrac{4}{43}$, multiply both fractions, top and bottom, by 10, and the fractions in between would be $\dfrac{31}{430}$, $\dfrac{32}{430}$, $\dfrac{33}{430}$, etc.

Answer 14:   E: $2 - \dfrac{3}{4} = \dfrac{5}{4}$. Its reciprocal is $\dfrac{4}{5}$.

Answer 15:   E: Let $100 be the original price. A 20% increase means a $20 increase. Then 20% decrease is ($120) × (.2) = $24. $120 − $24 = $96 or a $4 decrease. $\dfrac{\$4}{\$100}$ is a 4% decrease.

Answer 16:   E: Suppose we have $100. Increased by 20%, we have $120. But another 20% less (now on a larger amount) is $24 less. We are at $96, down 4%. Suppose we take 20% off first. We are at $80. Now 20% up (on a smaller amount) is $16. We are again at $96 or again down 4%.

Answer 17:   C: $100 discounted 50% is $50. 50% of $50 is $25, for a 75% discount. This is why after December, a 50% discount followed by 20% is 60%, but sounds like 70% off.

Answer 18:   **D:** If you memorized the fraction table, you would recognize this problem as the same as $\left(\frac{5}{6}\right)\left(\frac{3}{8}\right) = \frac{5}{16}$.

Answer 19:   **C:** It is easy if you've memorized the fraction table.

Answer 20:   **E:** If we changed each number to a fraction with a denominator of 48 (the LCM), we would get $\frac{19}{24} = \frac{38}{48}$, $\frac{1}{2} = \frac{24}{48}$, $\frac{3}{8} = \frac{18}{48}$, $\frac{3}{4} = \frac{36}{48}$, and $\frac{9}{16} = \frac{27}{48}$, and it is easy to see that $\frac{27}{48} = \frac{9}{16}$ is the middle number.

Answer 21:   **E:** There are two ways to do this problem. If you recognize that $\frac{31}{125} < \frac{1}{4}$ (= .25) since the bottom is more than 4 times the top, you see that choice E is the only one less than $\frac{1}{4}$. Another way to answer this question is to multiply the top and bottom of $\frac{31}{125}$ by 8, and get $\frac{248}{1000} = .248$, which is answer choice E.

Enough! Let's do some algebra now. We will see these topics throughout the book.

"The power of exponents will bring you strength and knowledge."

Exponents are a very popular topic on the GMAT. They are a good test of knowledge and thinking, are short to write, and it is relatively easy to make up new problems. Let's review some basic rules of exponents.

Rule	**Examples**
1. $x^m x^n = x^{m+n}$	$x^6 x^4 x = x^{11}$ and $(x^6 y^7)(x^4 y^{10}) = x^{10}y^{17}$
2. $\dfrac{x^m}{x^n} = x^{m-n}$ or $\dfrac{1}{x^{n-m}}$	$\dfrac{x^8}{x^6} = x^2$, $\dfrac{x^3}{x^7} = \dfrac{1}{x^4}$, and $\dfrac{x^4 y^5 z^9}{x^9 y^2 z^9} = \dfrac{y^3}{x^5}$
3. $(x^m)^n = x^{mn}$	$(x^5)^7 = x^{35}$
4. $(xy)^n = x^n y^n$	$(xy)^3 = x^3 y^3$ and $(x^7 y^3)^{10} = x^{70} y^{30}$
5. $\left(\dfrac{x}{y}\right)^n = \dfrac{x^n}{y^n}$	$\left(\dfrac{x}{y}\right)^6 = \dfrac{x^6}{y^6}$ and $\left(\dfrac{y^4}{z^5}\right)^3 = \dfrac{y^{12}}{z^{15}}$
6. $x^{-n} = \dfrac{1}{x^n}$ and $\dfrac{1}{x^{-m}} = x^m$	$2^{-3} = \dfrac{1}{2^3} = \dfrac{1}{8}$, $\dfrac{1}{4^{-3}} = 4^3 = 64$, $\dfrac{x^{-4}y^{-5}z^6}{x^{-6}y^4 z^{-1}} = \dfrac{x^6 z^6 z^1}{x^4 y^5 y^4} = \dfrac{x^2 z^7}{y^9}$, and $\left(\dfrac{x^3}{y^{-4}}\right)^{-2} = \left(\dfrac{y^{-4}}{x^3}\right)^2 = \dfrac{y^{-8}}{x^6} = \dfrac{1}{x^6 y^8}$
7. $x^0 = 1$, $x \neq 0$; 0^0 is indeterminate	$(7ab)^0 = 1$ and $7x^0 = 7(1) = 7$

8. $x^{\frac{p}{r}} = x^{\frac{\text{power}}{\text{root}}}$ Even though either order of computing the power and root give the same answer, we usually do the root first. $25^{\frac{3}{2}} = \left(\sqrt{25}\right)^3 = 125$; $4^{-\frac{3}{2}} = \dfrac{1}{4^{\frac{3}{2}}} = \dfrac{1}{\left(\sqrt{4}\right)^3} = \dfrac{1}{8}$

Because there are no calculators allowed in taking the GMAT, it might pay to remember the following:

$2^2 = 4$	$2^3 = 8$	$2^4 = 16$	$2^5 = 32$	$2^6 = 64$	$2^7 = 128$
$2^8 = 256$	$2^9 = 512$	$2^{10} = 1024$	$3^2 = 9$	$3^3 = 27$	$3^4 = 81$
$3^5 = 243$	$3^6 = 729$	$4^2 = 16$	$4^3 = 64$	$4^4 = 256$	$5^2 = 25$
$5^3 = 125$	$6^2 = 36$	$6^3 = 216$	$7^2 = 49$	$8^2 = 64$	$9^2 = 81$
$10^2 = 100$	$11^2 = 121$	$12^2 = 144$	$13^2 = 169$	$14^2 = 196$	$15^2 = 225$
$16^2 = 256$	$17^2 = 289$	$18^2 = 324$	$19^2 = 361$	$20^2 = 400$	$21^2 = 441$
$22^2 = 484$	$23^2 = 529$	$24^2 = 576$	$25^2 = 625$	$26^2 = 676$	$27^2 = 729$
$28^2 = 784$	$29^2 = 841$	$30^2 = 900$	$31^2 = 961$	$32^2 = 1024$	

Over the years, I have asked my students to memorize these powers for a number of reasons. In math, these powers occur often. They occur when we use the Pythagorean theorem (mostly the squares). They are used going backwards to find roots (next chapter). Also, they show number patterns of the squares (look at the last digits of all of them). All of these topics are on the GMAT.

Even though the GMAT does not ask comparison-type questions, like the GRE does and the SAT used to, it is necessary to be able to compare x and x^2. Similar results hold for x^3, x^4, and so on.

If $x > 1$, then $x^2 > x$, since $4^2 > 4$.

If $x = 1$, then $x^2 = x$, since $1^2 = 1$.

If $0 < x < 1$, then $x > x^2$, since $\dfrac{1}{2} > \left(\dfrac{1}{2}\right)^2 = \dfrac{1}{4}$!!

If $x = 0$, then $x = x^2$, since $0 = 0^2$.

If $x < 0$, then $x^2 > x$, since the square of a negative is a positive.

Also recall that if $0 < x < 1$, then $\dfrac{1}{x} > 1$, and if $x > 1$, then $0 < \dfrac{1}{x} < 1$.

Also if $-1 < x < 0$, then $\dfrac{1}{x} < -1$, and if $x < -1$, then $-1 < \dfrac{1}{x} < 0$.

Here are some exponential problems to practice, including some with negative exponents.

Example 1: Simplify the following:

Problem	Solution
a. $(-3a^4bc^6)(-5ab^7c^{10})(-100a^{100}b^{200}c^{2000}) =$	$-1500a^{105}b^{208}c^{2016}$
b. $(10ab^4c^7)^3 =$	$1000a^3b^{12}c^{21}$
c. $(4x^6)^2(10x)^3 =$	$16{,}000x^{15}$
d. $((2b^4)^3)^2 =$	$64b^{24}$
e. $(-b^6)^{101} =$	$-b^{606}$
f. $(-ab^8)^{202} =$	$a^{202}b^{1616}$
g. $\dfrac{24e^9f^7g^5}{72e^9f^{11}g^7} =$	$\dfrac{1}{3f^4g^2}$
h. $\dfrac{\left(x^4\right)^3}{x^4} =$	x^8
i. $\left(\dfrac{m^3n^4}{m^7n}\right)^5 =$	$\dfrac{n^{15}}{m^{20}}$
j. $\left(\dfrac{\left(p^4\right)^3}{\left(p^6\right)^5}\right)^{10} =$	$\dfrac{1}{p^{180}}$
k. $(-10a^{-4}b^5c^{-2})(4a^{-7}b^{-1}) =$	$\dfrac{-40b^4}{a^{11}c^2}$
l. $(3ab^{-3}c^4)^{-3} =$	$\dfrac{b^9}{27a^3c^{12}}$

m. $\left(3x^4\right)^{-4}\left(\left(\dfrac{1}{9x^8}\right)^{-1}\right)^2 =$ 1

n. $(2x^{-4})^2(3x^{-3})^{-2} =$ $\dfrac{4}{9x^2}$

o. $\left(\dfrac{(2y^3)^{-2}}{(4x^{-5})}\right)^{-2}$ $\dfrac{256y^{12}}{x^{10}}$

Q **Now, let's try some exercises.**

Exercise 1: $0 < x < 1$: Arrange in order of smallest to largest: x, x^2, x^3.

A. $x < x^2 < x^3$ D. $x^3 < x^2 < x$

B. $x < x^3 < x^2$ E. $x^3 < x < x^2$

C. $x^2 < x < x^3$

Exercise 2: $-1 < x < 0$: Arrange in order largest to smallest: x^2, x^3, x^4

A. $x^4 > x^3 > x^2$ D. $x^2 > x^3 > x^4$

B. $x^4 > x^2 > x^3$ E. $x^2 > x^4 > x^3$

C. $x^3 > x^4 > x^2$

Exercise 3: $0 < x < 1$.

I: $x > \dfrac{1}{x^2}$

II: $\dfrac{1}{x^2} > \dfrac{1}{x^4}$

III: $x - 1 > \dfrac{1}{x - 1}$

Which statement(s) are always true?

A. None   D. III only

B. I only   E. All

C. II only

Exercise 4: $(5ab^3)^3 =$

A. $15ab^6$ D. $125a^3b^6$

B. $75ab^6$ E. $125a^3b^9$

C. $125ab^9$

Exercise 5: $\dfrac{\left(2x^5\right)^3\left(3x^{10}\right)^2}{6x^{15}} =$

A. 1

D. $12x^{20}$

B. x^{20}

E. $12x^{210}$

C. $2x^{20}$

Exercise 6: $\left(\dfrac{12x^6}{24x^9}\right)^3 =$

A. $1728x^{27}$

D. $\dfrac{1}{8x^9}$

B. $\dfrac{1}{1728x^{27}}$

E. $\dfrac{1}{8x^{27}}$

C. $\dfrac{1}{6x^9}$

Exercise 7: $\dfrac{\left(4x^4\right)^3}{\left(8x^6\right)^2} =$

A. $\dfrac{1}{2}$

D. x^{28}

B. 1

E. $\dfrac{1}{2x}$

C. $\dfrac{1}{2}x^{28}$

Exercise 8: $-1 \le x \le 5$. Where is x^2 located?

A. $-1 \le x^2 \le 5$

D. $1 \le x^2 \le 10$

B. $0 \le x^2 \le 25$

E. $1 \le x^2 \le 25$

C. $1 \le x^2 \le 5$

Exercise 9: $2^m + 2^m =$

A. 2^{m+1}

D. 2^{m^2}

B. 2^{m+2}

E. 4^m

C. 2^{m+4}

Exercise 10: $\dfrac{m^{-5}n^6p^{-2}}{m^{-3}n^9p^0} =$

A. $m^2n^3p^2$

D. $\dfrac{1}{m^2n^3p^2}$

B. $\dfrac{1}{m^2n^3}$

E. None of these

C. $\dfrac{m^2}{n^3p^2}$

Exercise 11: If $8^{2n+1} = 2^{n+18}$; $n =$

A. 3

D. 13

B. 7

E. 17

C. 10

Exercise 12: $p = 4^n$; $4p =$

A. 4^{n+1}

D. 16^p

B. 4^{n+2}

E. 64^p

C. 3^{n+4}

Exercise 13: $y^3 = 64$; $y^{-2} =$

A. -4

D. $\dfrac{1}{16}$

B. $-\dfrac{1}{8}$

E. $-\dfrac{1}{16}$

C. $\dfrac{1}{8}$

Exercise 14: Suppose $x^2 = y^2$.

I: $x = y$.

II: $x = -y$.

III: $x^2 = xy$.

Which statements are always true?

A. None

D. III only

B. I only

E. All statements are true

C. II only

Exercise 15: $3^{-2} =$

A. $\dfrac{1}{3}$ D. $-\dfrac{1}{9}$

B. $\dfrac{1}{6}$ E. $-\dfrac{1}{6}$

C. $\dfrac{1}{9}$

Exercise 16: $x^{\frac{3}{4}} = 8; x =$

A. $\dfrac{32}{3}$ D. 256

B. 16   E. 1024

C. 64

Exercise 17: n is an integer, and $(-2)^{6n} = 8^{n+4}; n =$

A. 2   D. 6

B. 3   E. 8

C. 4

Ⓐ **Let's look at the answers.**

Answer 1:   D: If $0 < x < 1$, the higher the power, the smaller the number.

Answer 2:   E: Take, for example, $x = -\dfrac{1}{2}\ \left(-\dfrac{1}{2}\right)^2 = \dfrac{1}{4}; \left(-\dfrac{1}{2}\right)^3 = -\dfrac{1}{8};$ $\left(-\dfrac{1}{2}\right)^4 = \dfrac{1}{16}.$ Be careful! This exercise asks for largest to smallest. Notice that x^3 has to be the smallest because it is the only negative number, so the answer choices are reduced to two, B and E.

Answer 3:   D: Statement I is false: $0 < x < 1$, so $\dfrac{1}{x} > 1$ and $\dfrac{1}{x^2} > 1$ also. Statement II is false: $x^2 > x^4$, so $\dfrac{1}{x^2} < \dfrac{1}{x^4}$. Statement III is true: $0 < x < 1$; so $-1 < x - 1 < 0$; this means $\dfrac{1}{x-1} < -1$

Answer 4:   E: $5^3 a^3 (b^3)^3 = 125 a^3 b^9$.

Answer 5: D: $\left(\dfrac{8 \times 9}{6}\right) x^{15 + 20 - 15} = 12x^{20}$.

Answer 6: D: $\left(\dfrac{1}{2x^3}\right)^3 = \dfrac{1}{8x^9}$.

Answer 7: B: The numerator and denominator of the fraction each equal $64x^{12}$.

Answer 8: B: This is very tricky. Since 0 is between -1 and 5 and $0^2 = 0$.

Answer 9: A: This is one of the few truly hard problems because it is an addition problem:

$$2^m + 2^m = 1 \times 2^m + 1 \times 2^m = 2 \times 2^m = 2^1 \times 2^m = 2^{m+1}.$$

Similarly, $3^m + 3^m + 3^m = 3^{m+1}$ and four 4^m terms added equal 4^{m+1}.

Answer 10: D: $\dfrac{m^{-5}n^6p^{-2}}{m^{-3}n^9p^0} = \dfrac{m^3n^6}{m^5n^9p^2} = \dfrac{1}{m^2n^3p^2}$.

Answer 11: A: $8^{2n+1} = (2^3)^{2n+1} = 2^{n+18}$. If the bases are equal, the exponents must be equal. $3(2n+1) = n + 18$; so $n = 3$.

Answer 12: A: $p = 4^n$; $4p = 4(4^n) = 4^1 4^n = 4^{n+1}$.

Answer 13: D: $y^3 = 64$; $y = 4$; $y^{-2} = \left(\dfrac{1}{4}\right)^2 = \dfrac{1}{16}$.

Answer 14: A: None! Since $x = y$ or $x = -y$, all of the statements are true sometimes, but none is always true.

Answer 15: C: $3^{-2} = \left(\dfrac{1}{3}\right)^2 = \dfrac{1}{9}$.

Answer 16: B: $x^{\frac{3}{4}} = 8$; $x = \left(x^{\frac{3}{4}}\right)^{\frac{3}{4}} = 8^{\frac{4}{3}} = \left(\sqrt[3]{8}\right)^4 = 2^4 = 16$

Answer 17: C: We can ignore the minus sign, since both sides must be positive. $2^{6n} = 8^{n+4} = 2^{3(n+4)}$. So $6n = 3n + 12$, and $n = 4$.

Now let's go to a radical chapter.

CHAPTER 4: *Square Roots*

"*We must go to the root of the problem to be enlightened.*"

The square root symbol ($\sqrt{\ }$) is probably the one symbol most people actually like, even for people who don't like math. How else can you explain the square root symbol on a business calculator? I have yet to find a use for it. Here are some basic facts about square roots that you should know.

1. You should know the following square roots:

    $\sqrt{0} = 0$ $\qquad$ $\sqrt{1} = 1$ $\qquad$ $\sqrt{4} = 2$ $\qquad$ $\sqrt{9} = 3$ $\qquad$ $\sqrt{16} = 4$ $\qquad$ $\sqrt{25} = 5$

    $\sqrt{36} = 6$ $\qquad$ $\sqrt{49} = 7$ $\qquad$ $\sqrt{64} = 8$ $\qquad$ $\sqrt{81} = 9$ $\qquad$ $\sqrt{100} = 10$

    The numbers under the radicals (square root signs) are called perfect squares because their square roots are whole numbers.

2. $\sqrt{2} \approx 1.4$ (actually it is $1.414\ldots$), and $\sqrt{3} \approx 1.73$ (actually it is $1.732\ldots$, the year George Washington was born).

3. $\sqrt{\dfrac{a}{b}} = \dfrac{\sqrt{a}}{\sqrt{b}}$, so $\qquad$ $\sqrt{\dfrac{25}{9}} = \dfrac{5}{3}$ $\qquad$ $\sqrt{\dfrac{7}{36}} = \dfrac{\sqrt{7}}{6}$ $\qquad$ $\sqrt{\dfrac{45}{20}} = \sqrt{\dfrac{9}{4}} = \dfrac{3}{2}$

4. A method of simplification involves finding all the prime factors:

    $\sqrt{200} = \sqrt{(2)(2)(2)(5)(5)} = (2)(5)\sqrt{2} = 10\sqrt{2}$. We can also simplify by using $\sqrt{200} = \sqrt{100 \times 2} = 10\sqrt{2}$. With paper and pencil, the first method has always been better for my students. Without paper and pencil on the GMAT, you have to decide which method works best for you.

5. Adding and subtracting radicals involves combining like radicals:

    $4\sqrt{7} + 5\sqrt{11} + 6\sqrt{7} - 9\sqrt{11} = 10\sqrt{7} - 4\sqrt{11}$

6. Multiplication of radicals follows this rule: $a\sqrt{b} \times c\sqrt{d} = ac\sqrt{bd}$.

    Therefore, $3\sqrt{13} \times 10\sqrt{7} = 30\sqrt{91}$, and

    $10\sqrt{8} \times 3\sqrt{10} = 10 \times 3\sqrt{2 \times 2 \times 2 \times 2 \times 5} = 10 \times 3 \times 2 \times 2\sqrt{5} = 120\sqrt{5}$.

7. If a radical appears in the denominator of a fraction, rationalize the denominator by multiplying both numerator and denominator by the radical:

$$\frac{20}{7\sqrt{5}} = \frac{20}{7\sqrt{5}} \times \frac{\sqrt{5}}{\sqrt{5}} = \frac{20\sqrt{5}}{35} = \frac{4\sqrt{5}}{7} \text{ and } \frac{7}{\sqrt{45}} = \frac{7}{3\sqrt{5}} \times \frac{\sqrt{5}}{\sqrt{5}} = \frac{7\sqrt{5}}{15}$$

8. If $c, d > 0$, $\sqrt{c} + \sqrt{d} > \sqrt{c + d}$. Why? If we square the right side, we get $c + d$. If we square the left side, we get $c + d +$ the middle term ($2\sqrt{cd}$).

9. You should know the following cube roots:

$$\sqrt[3]{1} = 1 \qquad \sqrt[3]{8} = 2 \qquad \sqrt[3]{27} = 3 \qquad \sqrt[3]{64} = 4 \qquad \sqrt[3]{125} = 5$$

$$\sqrt[3]{-1} = -1 \qquad \sqrt[3]{-8} = -2 \qquad \sqrt[3]{-27} = -3 \qquad \sqrt[3]{-64} = -4 \qquad \sqrt[3]{-125} = -5$$

10. The square root or any even root of a negative number is imaginary (undefined) and not on the GMAT.

11. The cube root or any odd root of a positive is a positive, a negative is a negative, and 0 is 0. All are always defined.

12. The square root varies according to the value of the radicand:

If $a > 1$, $a > \sqrt{a}$. For example, $9 > \sqrt{9}$.

If $a = 1$, $a = \sqrt{a}$ since the square root of 1 is 1.

If $0 < a < 1$, $a < \sqrt{a}$. When we take the square root of a positive number, it becomes closer to 1, so $\sqrt{\frac{1}{4}} = \frac{1}{2} > \frac{1}{4}$.

If $a = 0$, $a = \sqrt{a}$ since the square root of 0 is 0.

If $a > 0$ and $\sqrt{a} < 1$, then $\frac{1}{\sqrt{a}} > 1$. Also, if $\sqrt{a} > 1$, then $\frac{1}{\sqrt{a}} < 1$.

Note $\sqrt{9} = 3$, $-\sqrt{9} = -3$, but $\sqrt{-9}$ is imaginary. The equation $x^2 = 9$ has two solutions, $\pm\sqrt{9}$, or ± 3, which stands for both $+3$ and -3.

The GMAT uses square roots often in comparison problems.

Q   **Let's try some exercises.**

Exercise 1: $\left(\sqrt{12} + \sqrt{27}\right)^2 =$

A. 15   D. 225

B. 39   E. 675

C. 75

Exercise 2:   Suppose $0 < a < 1$.

I: $a^2 > \sqrt{a}$

II: $\sqrt{a} > \sqrt{a^3}$

III: $\sqrt{a} > \dfrac{1}{\sqrt{a^7}}$

A. I is correct

B. II is correct

C. III is correct

D. I and III are correct

E. II and III are correct

Exercise 3: $c = \left(\dfrac{1}{17}\right)^2 - \sqrt{\dfrac{1}{17}}$. Which answer choice is true for c?

A. $c < -2$

B. $-2 < c < -1$

C. $-1 < c < 0$

D. $0 < c < 1$

E. $1 < c < 2$

Exercise 4: $0 < m < 1$. Arrange in order, smallest to largest, $a = \dfrac{1}{m}$, $b = \dfrac{1}{m^2}$, $c = \dfrac{1}{\sqrt{m}}$.

A. $a < b < c$

B. $a < c < b$

C. $b < c < a$

D. $b < a < c$

E. $c < a < b$

Exercise 5:   If $\sqrt[3]{-87} = x$:

A. $-10 < x < -9$

B. $-9 < x < -8$

C. $-5 < x < -4$

D. $-4 < x < -3$

E. x is undefined

A **Let's look at the answers.**

Answer 1:   **C:** $\sqrt{12} = \sqrt{2 \times 2 \times 3} = 2\sqrt{3}$ and $\sqrt{27} = \sqrt{3 \times 3 \times 3} = 3\sqrt{3}$. Adding, we get $5\sqrt{3}$. Squaring, we get $25\sqrt{9} = 25 \times 3 = 75$.

Answer 2:   **B:** Let's look at the statements one by one.

Statement I: If we square a number between 0 and 1, we make it closer to 0. If we take the square root of the same number, we make it closer to 1. Statement I is wrong.

Statement II: From the previous chapter, if $0 < a < 1$, $a > a^3$. So are its square roots. So statement II is true.

Statement III: $\sqrt{a} < 1$. $a^7 < 1$. So $\sqrt{a^7} < 1$. Then $\dfrac{1}{\sqrt{a^7}} > 1$. Statement III is false.

Answer 3:   **C:** $\dfrac{1}{17}$ squared is less than $\dfrac{1}{17}$, the square root is more than $\dfrac{1}{17}$, and both numbers are between 0 and 1. Subtracting the larger from the smaller, the answer must be C.

Answer 4:   **E:** If we take $m = \dfrac{1}{4}$, we see that $\sqrt{m} > m > m^2$. That makes $\dfrac{1}{\sqrt{m}} < \dfrac{1}{m} < \dfrac{1}{m^2}$, or $c < a < b$.

Answer 5:   **C:** Since $\sqrt[3]{-125} = -5$ and $\sqrt[3]{-64} = -4$, and since -87 is between -125 and -64, the cube root of -87 must be between -5 and -4.

We'll see more root problems in the later chapters and in the review.

"Along our journey, we must learn to do. It will help us become truly happy."

Algebraic manipulative skills such as those in this chapter are areas that high school courses have tended to de-emphasize since 1985. It is necessary to show you how to do these problems and give you extra problems to practice. Of course, included will be the kind of questions the GMAT asks.

COMBINING LIKE TERMS

Like terms are terms with the same letter combination (or no letter). The same letter must also have the same exponents.

Example 1: Are the following terms like or unlike:

  a. $4x$ and $-5x$

  b. $4x$ and $4x^2$

  c. xy^2 and x^2y

Solutions: **a.** $4x$ and $-5x$ are like terms even though their numerical coefficients are different.

  b. $4x$ and $4x^2$ are unlike terms

  c. xy^2 and x^2y are unlike; $xy^2 = xyy$ and $x^2y = xxy$.

Combining like terms means adding or subtracting their numerical coefficients; exponents are unchanged. Unlike terms cannot be combined.

Example 2:   Simplify:

Problem	Solution
a. $3m + 4m + m =$	$8m$
b. $8m + 2n + 7m - 7n =$	$15m - 5n$
c. $3x^2 + 4x - 5 - 7x^2 - 4x + 8 =$	$-4x^2 + 3$

DISTRIBUTIVE LAW

The **Distributive Law** states:

$$a(x + y) = ax + ay$$

Example 3:   Perform the indicated operations:

Problem	Solution
a. $4(3x - 7) =$	$12x - 28$
b. $5(2a - 5b + 3c) =$	$10a - 25b + 15c$
c. $3x^4(7x^3 - 4x - 1) =$	$21x^7 - 12x^5 - 3x^4$
d. $4(3x - 7) - 5(4x - 2) =$	$12x - 28 - 20x + 10 = -8x - 18$

BINOMIAL PRODUCTS

A **binomial** is a two-term expression, such as $x + 2$. We use the **FOIL method** to multiply a binomial by a binomial. FOIL is an acronym for First, Outer, Inner, Last. This means to multiply the first two terms, then the outer terms, then the inner terms, and finally the last two terms.

Example 4:   Multiply $(x + 4)(x + 6)$

Solution:

Multiplying, we get $x^2 + 6x + 4x + 24 = x^2 + 10x + 24$.

Example 5:   Perform the indicated multiplications:

Problem	Solution

a. $(x + 7)(x + 4) =$ $x^2 + 4x + 7x + 28 = x^2 + 11x + 28$

b. $(x - 5)(x - 2) =$ $x^2 - 7x + 10$

c. $(x + 6)(x - 3) =$ $x^2 + 3x - 18$

d. $(x + 6)(x - 8) =$ $x^2 - 2x - 48$

e. $(x + 5)(x - 5) =$ $x^2 - 5x + 5x - 25 = x^2 - 25$

f. $(x + 5)^2 =$ $(x + 5)(x + 5) = x^2 + 10x + 25$

g. $(x - 10)^2 =$ $x^2 - 20x + 100$

h. $(2x + 5)(3x - 10) =$ $6x^2 - 5x - 50$

i. $3(x + 4)(x + 5) =$ $3(x^2 + 9x + 20) = 3x^2 + 27x + 60$

j. $7(4x + 3)(4x - 3) =$ $7(16x^2 - 9) = 112x^2 - 63$

You should know the following common binomial products:

$$(a + b)(a - b) = a^2 - b^2$$

$$(a - b)(a - b) = a^2 - 2ab + b^2$$

$$(a + b)(a + b) = a^2 + 2ab + b^2$$

 For a perfect square $(a + b)^2$, the first term of the resulting trinomial is the first term squared (a^2), and the third term of the resulting trinomial is the last term squared (b^2). The middle term is twice the product of the two terms of the binomial (2ab), so

$$(a + b)^2 = a^2 + 2ab + b^2$$

Example 6:   Perform the indicated multiplications:

Problem	Solution
a. $(a + 4)(a + 7)$	$a^2 + 11a + 28$
b. $(b + 5)(b + 6)$	$b^2 + 11b + 30$
c. $(c + 1)(c + 9)$	$c^2 + 10c + 9$
d. $(d + 4)(d + 8)$	$d^2 + 12d + 32$
e. $(e + 11)(e + 10)$	$e^2 + 21e + 110$
f. $(f - 6)(f - 2)$	$f^2 - 8f + 12$
g. $(g - 10)(g - 20)$	$g^2 - 30g + 200$
h. $(h - 4)(h - 3)$	$h^2 - 7h + 12$
i. $(i - 1)(i - 7)$	$i^2 - 8i + 7$
j. $(j - 3)(j - 5)$	$j^2 - 8j + 15$
k. $(k + 5)(k - 2)$	$k^2 + 3k - 10$
l. $(m + 5)(m - 8)$	$m^2 - 3m - 40$
m. $(n - 6)(n + 2)$	$n^2 - 4n - 12$
n. $(p - 8)(p + 10)$	$p^2 + 2p - 80$
o. $(q - 5r)(q + 2r)$	$q^2 - 3qr - 10r^2$
p. $(s + 3)^2$	$s^2 + 6s + 9$
q. $(t - 4)^2$	$t^2 - 8t + 16$
r. $(3u + 5)^2$	$9u^2 + 30u + 25$
s. $(5v - 4)^2$	$25v^2 - 40v + 16$

t. $(ax + by)^2$ $\qquad\qquad$ $a^2x^2 + 2abxy + b^2y^2$

u. $(be - ma)^2$ $\qquad\qquad$ $b^2e^2 - 2\,beam + m^2a^2$

v. $(w + x)\,(w - x)$ $\qquad\qquad$ $w^2 - x^2$

w. $(a - 11)\,(a + 11)$ $\qquad\qquad$ $a^2 - 121$

x. $(am - 7)\,(am + 7)$ $\qquad\qquad$ $a^2m^2 - 49$

y. $(a^2b + c)\,(a^2b - c)$ $\qquad\qquad$ $a^4b^2 - c^2$

z. $3(x + 5)\,(x - 2)$ $\qquad\qquad$ $3x^2 + 9x - 30$

aa. $-4(2x - 5)\,(3x - 4)$ $\qquad\qquad$ $-24x^2 + 92x - 80$

bb. $x(2x - 5)\,(4x + 7)$ $\qquad\qquad$ $8x^3 - 6x^2 - 35x$

cc. $5(x - 5)\,(x + 5)$ $\qquad\qquad$ $5x^2 - 125$

Ⓠ Let's do some exercises.

Exercise 1: $x^2 - y^2 = 24;\ 3(x + y)\,(x - y) =$

A. 8 $\qquad\qquad$ D. 72

B. 24 $\qquad\qquad$ E. 13,824

C. 27

Exercise 2: $x + y = m;\ x - y = \dfrac{1}{m};\ x^2 - y^2 =$

A. m^2 $\qquad\qquad$ D. $\dfrac{1}{m}$

B. m $\qquad\qquad$ E. $\dfrac{1}{m^2}$

C. 1

Exercise 3: $\left(x + \dfrac{1}{x}\right)^2 = 64; \ x^2 + \dfrac{1}{x^2} =$

A. 9   D. 65

B. 62   E. 66

C. 64

Exercise 4: $x^2 + y^2 = 20$ and $xy = -6$. Then $(x + y)^2 =$

A. 8   D. 26

B. 14   E. 32

C. 20

Exercise 5: $(x - y)^2 + 4xy =$

A. $x^2 + 8x + y^2$ D. $(x + y)^2$

B. $x^2 + 4x + y^2$ E. $x^2 + y^2$

C. $x^2 - y^2$

Ⓐ **Let's look at the answers.**

Answer 1:   D: $3(x + y)(x - y) = 3(x^2 - y^2) = 3(24) = 72$.

Answer 2:   C: $(x + y)(x - y) = x^2 - y^2 = \dfrac{m}{1} \times \dfrac{1}{m} = 1$.

Answer 3:   B: $\left(x + \dfrac{1}{x}\right)\left(x + \dfrac{1}{x}\right) = x^2 + 2(x)\left(\dfrac{1}{x}\right) + \dfrac{1}{x^2} = x^2 + \dfrac{1}{x^2} + 2 = 64$. So

$x^2 + \dfrac{1}{x^2} = 64 - 2 = 62$.

Answer 4:   A: $(x + y)^2 = x^2 + 2xy + y^2 = x^2 + y^2 + 2xy = 20 + 2(-6) = 8$.

Answer 5:   D: $x^2 - 2xy + y^2 + 4xy = x^2 + 2xy + y^2 = (x + y)^2$.

Let's go on to factoring.

FACTORING

Factoring is the reverse of the distributive law. There are three types of factoring you need to know: largest common factor, difference of two squares, and trinomial factorization.

If the distributive law says $x(y + z) = xy + xz$, then taking out the largest common factor says $xy + xy = x(y + z)$. Let's demonstrate a few factoring examples.

Example 7: Factor:

Problem	Answer	Explanation
a. $4x + 6y - 8$	$2(2x + 3y - 4)$	2 is the largest common factor.
b. $8ax + 12ay - 40az$	$4a(2x + 3y - 10z)$	4 is the largest common factor; a is also a common factor
c. $10a^4y^6z^3 - 15a^7y$	$5a^4y(2y^5 z^3 - 3a^3)$	The largest common factor and the lowest power of each common variable is factored out. a^4 and y; but not z because it is not in both terms.
d. $x^4y - xy^3 + xy$	$xy(x^3 - y^2 + 1)$	Factor out the lowest power of each common variable. Three terms in the original give three terms in parentheses. Note that $1 \times xy = xy$.
e. $9by + 12be + 4ye$	prime	Some expressions cannot be factored.

Difference of Two Squares

Since $(a + b)(a - b) = a^2 - b^2$, factoring tells us that $a^2 - b^2 = (a + b)(a - b)$.

Example 8:

Problem	Answer	Explanation
a. $x^2 - 25$	$(x + 5)(x - 5)$ or $(x - 5)(x + 5)$	Either order is OK.
b. $x^2 - 121$	$(x + 11)(x - 11)$	
c. $9a^2 - 25b^2$	$(3a + 5b)(3a - 5b)$	

d. $5a^3 - 20a$ $5a(a^2 - 4) = 5a(a + 2)(a - 2)$ Factor out largest common factor first, then use the difference of two squares.

e. $x^4 - y^4$ $(x^2 + y^2)(x^2 - y^2) =$
$(x^2 + y^2)(x + y)(x - y)$ This is the difference of two squares where the square roots in the factors are also squares. Sum of two squares doesn't factor, but use the difference of two squares again.

Example 9:   Factor completely:

Problem   Solution

a. $12am + 18an$ $6a(2m + 3n)$

b. $6at - 18st + 4as$ $2(3at - 9st + 2as)$

c. $10ax + 15ae - 16ex$ Prime

d. $18a^5c^6 - 27a^3c^8$ $9a^3c^6(2a^2 - 3c^2)$

e. $25a^4b^7c^9 - 75a^8b^9c^{10}$ $25a^4b^7c^9(1 - 3a^4b^2c)$

f. $a^4b^5 + a^7b - ab$ $ab(a^3b^4 + a^6 - 1)$

g. $9 - x^2$ $(3 + x)(3 - x)$

h. $x^4 - 36y^2$ $(x^2 + 6y)(x^2 - 6y)$

i. $2x^3 - 98x$ $2x(x + 7)(x - 7)$

j. $a^4 - 81b^2$ $(a^2 + 9b)(a^2 - 9b)$

k. $x^2 - 49$ $(x + 7)(x - 7)$

l. $5z^2 - 25$ $5(z^2 - 5)$

m. $a^4 - c^8$ $(a^2 + c^4)(a + c^2)(a - c^2)$

n. $2a^9 - 32a$ $2a(a^4 + 4)(a^2 + 2)(a^2 - 2)$

Q   **Let's do some exercises.**

Exercise 6: $8x + 6y = 30$. Then $20x + 15y =$

A. 15   D. 75

B. 30   E. 150

C. 60

Exercise 7: $x^2 - 4 = 47 \times 43$; $x =$

A. 41   D. 47

B. 43   E. 49

C. 45

A   **Let's look at the answers.**

Answer 6:   D: $8x + 6y = 2(4x + 3y) = 30$; so $4x + 3y = 15$. $20x + 15y = 5(4x + 3y)$
$= 5 \times 15 = 75$.

Answer 7:   C: $x^2 - 4 = (x + 2)(x - 2) = 47 \times 43$, or $(45 + 2)(45 - 2)$. So $x = 45$.

Later in this chapter and in later chapters, we will see more GMAT comparison questions. For now, let's do trinomial factoring.

Factoring Trinomials

Factoring trinomials is a puzzle, a game, which is rarely done well in high school and even more rarely practiced. Let's learn the factoring game.

First, let's rewrite the first four parts of Example 5 backward and look at them.

a. $x^2 + 11x + 28 = (x + 7)(x + 4)$

b. $x^2 - 7x + 10 = (x - 5)(x - 2)$

c. $x^2 + 3x - 18 = (x + 6)(x - 3)$

d. $x^2 - 2x - 48 = (x + 6)(x - 8)$

Each term starts with $x^2 (= +1x^2)$, so the first sign is $+$. We'll call the sign in front of the x term the middle sign, and we'll call the sign in front of the number term the last sign.

Let's look at **a** and **b** above to state some rules of the game:

1. If the last sign (in the trinomial) is +, then both signs (in the parentheses) must be the same. The reason? $(+) \times (+) = +$, and $(-) \times (-) = +$.

2. Only if the last sign is +, look at the sign of the middle term. If it is +, both factors have a + sign (as in A); if it is −, both factors have a − sign (as in B).

3. If the last sign is −, the signs in the two factors must be different.

Now, let's play the game,

Example 10: Factor: $x^2 - 16x + 15$.

Solution: 1. The last sign + means both signs are the same. The middle sign − means both are −.

2. The only factors of x^2 are $(x)(x)$. Look at the number term 15. The factors of 15 are (3)(5) and (1)(15). So $(x - 5) (x - 3)$ and $(x - 15)$ $(x - 1)$ are the only possibilities. We have chosen the first and last terms to be correct, so we only do the middle term. The first, $-8x$, is wrong; the second, $-16x$, is correct.

3. The answer is $(x - 15) (x - 1)$. If neither worked, the trinomial couldn't be factored.

Example 11: Factor completely: $x^2 - 4x - 21$.

Solution: 1. The last sign negative means the signs of the factors are different.

2. $x^2 = x(x)$ and the factors of 21 are 7 and 3 or 21 and 1. We want the pair that totals the middle number (−4), so −7 and 3 are correct. Note that the larger factor (7) gets the minus sign because the middle number is negative.

3. The answer is $(x - 7)(x + 3)$.

Note   *If you multiply the inner and outer terms and get the right number but the wrong sign, both signs in the parentheses must be changed.*

The game gets more complicated if the coefficient of x^2 is not 1.

Example 12: Factor completely: $4x^2 + 4x - 15$.

Solution:
1. The last sign is $-$, so the signs of the factors must be different.
2. The factors of $4x^2$ are $(4x)(x)$ or $(2x)(2x)$.
3. The factors of 15 are 3 and 5, or 1 and 15. Let's write out all the possibilities and the resulting middle terms. We are looking for terms whose difference is $4x$.

   $(4x \quad 3)(x \quad 5)$; middle terms are $3x$ and $20x$, no way to get $4x$.

   $(4x \quad 5)(x \quad 3)$; middle terms are $5x$ and $12x$, and again, there is no way to get $4x$.

   $(4x \quad 15)(x \quad 1)$; middle terms are $15x$ and $4x$, wrong!

   $(4x \quad 1)(x \quad 15)$; middle terms are $1x$ and $60x$, the next county.

   $(2x \quad 1)(2x \quad 15)$; middle terms are $2x$ and $30x$, wrong again!

   $(2x \quad 3) (2x \quad 5)$; middle terms are $6x$ and $10x$, correct, whew!
4. The minus sign goes in front of the 3 and the plus sign goes in front of the 5, so the answer is $(2x - 3)(2x + 5)$.

Example 13: Factor completely: $3x^2 + 15x + 12$.

Solution: Take out the common factor first.

$$3x^2 + 15x + 12 = 3(x^2 + 5x + 4) = 3(x + 4)(x + 1).$$

Example 14. Factor completely; coefficients may be integers only.

Problem	Solution
a. $x^2 + 11x + 24$	$(x + 3)(x + 8)$
b. $x^2 - 11x - 12$	$(x - 12)(x + 1)$
c. $x^2 + 5x - 6$	$(x + 6)(x - 1)$
d. $x^2 - 20x + 100$	$(x - 10)^2$
e. $x^2 - x - 2$	$(x - 2)(x + 1)$
f. $x^2 - 15x + 56$	$(x - 7)(x - 8)$
g. $x^2 + 8x + 16$	$(x + 4)^2$

h. $x^2 - 6x - 16$ $(x - 8)(x + 2)$

i. $x^2 - 17x + 42$ $(x - 14)(x - 3)$

j. $x^2 + 5xy + 6y^2$ $(x + 2y)(x + 3y)$

k. $3x^2 - 6x - 9$ $3(x - 3)(x + 1)$

l. $4x^2 + 16x - 20$ $4(x + 5)(x - 1)$

m. $x^3 - 12x^2 + 35x$ $x(x - 7)(x - 5)$

n. $2x^8 + 8x^7 + 6x^6$ $2x^6(x + 3)(x + 1)$

o. $x^4 - 10x^2 + 9$ $(x + 3)(x - 3)(x + 1)(x - 1)$

p. $x^4 - 8x^2 - 9$ $(x^2 + 1)(x + 3)(x - 3)$

q. $2x^2 - 5x + 3$ $(2x - 3)(x - 1)$

r. $2x^2 + 5x - 3$ $(2x - 1)(x + 3)$

s. $5x^2 - 11x + 2$ $(5x - 1)(x - 2)$

t. $9x^2 + 21x - 8$ $(3x + 8)(3x - 1)$

u. $3x^2 - 8x - 3$ $(3x + 1)(x - 3)$

v. $6x^2 - 13x + 6$ $(3x - 2)(2x - 3)$

w. $6x^2 + 35x - 6$ $(6x - 1)(x + 6)$

x. $9x^2 + 71x - 8$ $(9x - 1)(x + 8)$

y. $9x^4 + 24x^3 + 12x^2$ $3x^2(3x + 2)(x + 2)$

Q **Let's do a couple of exercises.**

Exercise 8:   If $x^2 + 5x + 4 = 27$, then $3(x + 1)(x + 4) =$

A. 9   D. 81

B. 24   E. 243

C. 30

Exercise 9:   If $(x - 6)$ is a factor of $x^2 + kx - 48$, $k =$

A. -288 D. 2

B. -14 E. 14

C. -2

A **Let's look at the answers.**

Answer 8:   D: $x^2 + 5x + 4 = (x + 1)(x + 4) = 27$; $3(x + 1)(x + 4) = 3(27) = 81$.

Answer 9:   D: $(-6)(?) = -48$; $? = 8$; $8 + (-6) = 2 = k$. To check, $(x - 6)(x + 8) = x^2 + 2x - 48$.

Again, later on, we will have more problems involving trinomial factoring.

ALGEBRAIC FRACTIONS

Except for adding and subtracting, the techniques for algebraic fractions are easy to understand. They must be practiced, however.

Reducing Fractions

Factor the top and bottom; cancel factors that are the same.

Example 15: Reduce the following fractions:

Problem   Solution

a. $\dfrac{x^2 - 9}{x^2 - 3x}$ $\dfrac{(x + 3)(x - 3)}{x(x - 3)} = \dfrac{x + 3}{x}$

b. $\dfrac{2x^3 + 10x^2 + 8x}{x^4 + x^3}$ $\dfrac{2x(x + 4)(x + 1)}{x^3(x + 1)} = \dfrac{2(x + 4)}{x^2}$

c. $\dfrac{x - 9}{9 - x}$ $\dfrac{(x - 9)}{-1(x - 9)} = -1$

Q **Let's do a few more exercises.**

Exercise 10: $\dfrac{16x + 4}{4} =$

A. $4x$ D. $12x + 1$

B. $4x + 1$ E. $16x + 1$

C. $12x$

Exercise 11: $\dfrac{x^2 - 16}{x - 4} =$

A. $x + 4$ D. $2x - 4$

B. $x - 4$ E. $x^2 - x - 12$

C. $2x + 4$

Exercise 12: $\dfrac{2x^2 + 10x + 12}{2x + 6} =$

A. x D. $11x + 2$

B. $x + 1$ E. $11x + 6$

C. $x + 2$

A **Let's look at the answers.**

Answer 10: B: $\dfrac{16x + 4}{4} = \dfrac{4(4x + 1)}{4} = 4x + 1$. Another way to do this problem is

$\dfrac{16x + 4}{4} = \dfrac{16x}{4} + \dfrac{4}{4} = 4x + 1.$

Answer 11:   A: $\dfrac{(x+4)(x-4)}{x-4} = x+4.$

Answer 12:   C: $\dfrac{2(x+2)(x+3)}{2(x+3)} = \dfrac{2(x+2)}{2} = x+2.$

Multiplication and Division of Fractions

Algebraic fractions use the same principle as multiplication and division of numerical fractions except we factor all tops and bottoms, canceling one factor in any top with its equivalent in any bottom. In a division problem, we must remember to invert the second fraction first and then multiply.

Example 16: $\dfrac{x^2-25}{(x+5)^3} \times \dfrac{x^3+x^2}{x^2-4x-5} =$

Solution: $\dfrac{(x+5)(x-5)}{(x+5)(x+5)(x+5)} \times \dfrac{x^2(x+1)}{(x-5)(x+1)} = \dfrac{x^2}{(x+5)^2}$

Example 17: $\dfrac{x^4+4x^2}{x^6} \div \dfrac{x^4-16}{x^2+3x-10} =$

Solution: $\dfrac{x^2(x^2+4)}{x^6} \times \dfrac{(x+5)(x-2)}{(x^2+4)(x-2)(x+2)} = \dfrac{x+5}{x^4(x+2)}$

Adding and Subtracting Algebraic Fractions

It might be time to review the section in Chapter 2 on adding and subtracting fractions. Follow these steps:

1. If the bottoms are the same, add the tops, reducing if necessary.

2. If the bottoms are different, factor the denominators.

3. The LCD is the product of the most number of times a prime appears in any one denominator.

4. Multiply top and bottom by "what's missing."

5. Add (subtract) and simplify the numerators; reduce, if possible.

Example 18: $\dfrac{x}{36-x^2} - \dfrac{6}{36-x^2} =$

Solution: $\dfrac{x-6}{36-x^2} = \dfrac{x-6}{(6-x)(6+x)} = \dfrac{-1}{x+6}$

Example 19: $\dfrac{5}{12xy^3} + \dfrac{9}{8x^2y} =$

Solution: $\dfrac{5}{2 \times 2 \times 3xyyy} + \dfrac{9}{2 \times 2 \times 2xxy}$

$= \dfrac{5(2x)}{2 \times 2 \times 2 \times 3xxyyy} + \dfrac{9(3yy)}{2 \times 2 \times 2 \times 3xxyyy} = \dfrac{10x + 27y^2}{24x^2y^3}$

Example 20: $\dfrac{2}{x^2 + 4x + 4} + \dfrac{3}{x^2 + 5x + 6} =$

Solution: $\dfrac{2}{(x + 2)(x + 2)} + \dfrac{3}{(x + 2)(x + 3)}$

$= \dfrac{2(x + 3)}{(x + 2)(x + 2)(x + 3)} + \dfrac{3(x + 2)}{(x + 2)(x + 3)(x + 2)} = \dfrac{5x + 12}{(x + 2)(x + 2)(x + 3)}$

This section has been included in the book for completeness. However, since this is a computer-adaptive test (CAT), questions like this should be few and far between.

Simplifying Complex Fractions

The GMAT seem to like questions about complex fractions, but mostly with numbers. Again you must do these without paper and pencil. With a little practice, you can!!!!

Example 21: Simplify $\dfrac{2 - \frac{5}{6}}{\frac{2}{3} + \frac{7}{8}}$.

Solution: Find the LCD of all the terms (24) and multiply each term by it.

$$\dfrac{\frac{2}{1} \times \frac{24}{1} - \frac{5}{6} \times \frac{24}{1}}{\frac{2}{3} \times \frac{24}{1} + \frac{7}{8} \times \frac{24}{1}} = \dfrac{48 - 20}{16 + 21} = \dfrac{28}{37}$$

Note *When you multiply each term by 24, all the fractions disappear except for the major one.*

Example 22: Simplify $\dfrac{1 - \frac{25}{x^2}}{1 - \frac{10}{x} + \frac{25}{x^2}}$.

Solution: The LCD is x^2.

$$\dfrac{x^2 \left(1 - \frac{25}{x^2}\right)}{x^2 \left(1 - \frac{10}{x} + \frac{25}{x^2}\right)} = \dfrac{x^2 - 25}{x^2 - 10x + 25} = \dfrac{(x + 5)(x - 5)}{(x - 5)(x - 5)} = \dfrac{x + 5}{(x - 5)}$$

Q **Let's do some more exercises.**

Exercise 13: The reciprocal of $2 - \dfrac{1}{4}$ is

A. $\dfrac{1}{2} - 4$ D. $\dfrac{4}{9}$

B. $-\dfrac{4}{7}$ E. $\dfrac{4}{7}$

C. $-\dfrac{4}{9}$

Exercise 14: $y = \dfrac{1}{x}$. Then $\dfrac{1-x}{1-y} =$

A. $-x$ D. $(1-x)$

B. x E. $(x-1)$

C. 1

Exercise 15: $\dfrac{2\frac{1}{5} + \frac{3}{5}}{2\frac{1}{5} - \frac{3}{5}} =$

A. -1 D. $\dfrac{7}{4}$

B. 1 E. $\dfrac{14}{5}$

C. $\dfrac{7}{5}$

A **Let's look at the answers.**

Answer 13: E: $2 - \dfrac{1}{4} = \dfrac{7}{4}$. Its reciprocal is $\dfrac{4}{7}$.

Answer 14: A: $\dfrac{1-x}{1-y} = \dfrac{1-x}{1-\frac{1}{x}} = \dfrac{x(1-x)}{x(1-\frac{1}{x})} = \dfrac{x(1-x)}{x-1} = -x$. The answer is A since

$\dfrac{1-x}{x-1} = -1$.

Answer 15:   **D:** The top is $\dfrac{11}{5} + \dfrac{3}{5} = \dfrac{14}{5}$; the bottom is $\dfrac{11}{5} - \dfrac{3}{5} = \dfrac{8}{5}$. So the fraction simplifies to $\dfrac{\frac{14}{5}}{\frac{8}{5}}$; the 5's cancel, and $\dfrac{14}{8} = \dfrac{7}{4}$.

Let's go on to equations.

CHAPTER 6: *First-Degree and Quadratic Equations and Inequalities*

"The equations here are the equations for life. Master them, and it will bring you joy."

FIRST-DEGREE EQUATIONS

In high school, the topic of first-degree equations was probably the most popular of all. The GMAT asks questions that are usually not too long and usually not too tricky. To review, here are the steps to solving first-degree equations. If you get good at these, you will know when to use the steps in another order.

To solve for *x*, follow these steps:

1. Multiply by the LCD to get rid of fractions. Cross-multiply if there are only two fractions.

2. If the "*x*" term appears only on the right, switch the sides.

3. Multiply out all parentheses by using the distributive law.

4. On each side, combine like terms.

5. Add the opposite of the *x* term on the right to each side.

6. Add the opposite of the non-*x* term(s) on the left to each side.

7. Factor out the *x*. This step occurs only if there is more than one letter in a problem.

8. Divide each side by the whole coefficient of *x*, including the sign.

 *The **opposite** of a term is the same term with its opposite sign. So the opposite of 3x is −3x, the opposite of −7y is + 7y, and the opposite of 0 is 0. The technical name for "opposite" is **additive inverse**.*

Believe it or not, it took a long time to get the phrasing of this list just right.

Example 1:   Solve for x: $7x - 2 = 10x + 13$

Solution: Steps 1–4 are not present.

$7x - 2 = 10x + 13$ Step 5: Add $-10x$ to each side.

$-3x - 2 = +13$ Step 6: Add $+2$ to each side.

$-3x = 15$ Step 8: Divide each side by -3.

$x = -5$ Solution

Example 2:   Solve for x: $7 = 2(3x - 5) - 4(x - 6)$

Solution: $7 = 2(3x - 5) - 4(x - 6)$ No Step 1. Step 2: Switch sides.

$2(3x - 5) - 4(x - 6) = 7$ Step 3: Multiply out the parentheses.

$6x - 10 - 4x + 24 = 7$ Step 4: Combine like terms on

                                       each side.

$2x + 14 = 7$ No Step 5. Step 6: Add -14 to

                                       each side.

$2x = -7$ No Step 7. Step 8: Divide each side by 2.

$x = -\dfrac{7}{2}$ The answer doesn't have to be an

                                       integer.

Example 3:   Solve for x: $\dfrac{x}{4} + \dfrac{x}{6} = 1$

Solution: $\dfrac{x}{4} + \dfrac{x}{6} = 1$ Step 1: Multiply each term by 12

$3x + 2x = 12$ Step 4: Combine like terms

$5x = 12$ Step 8: Divide each side by 5

$x = \dfrac{12}{5}$ Solution

Example 4: Solve for x: $y = \dfrac{3x - 5}{x - 7}$

Solution: Write $y = \dfrac{y}{1}$: $\dfrac{y}{1} = \dfrac{3x - 5}{x - 7}$ Step 1: Cross-multiply

$(x - 7)y = 1(3x - 5)$ Step 3: Distribute

$xy - 7y = 3x - 5$ Step 5: Add $-3x$ to each side

$xy - 3x - 7y = -5$ Step 6: Add $7y$ to each side

$xy - 3x = 7y - 5$ Step 7: Factor out the x from the left

$x(y - 3) = 7y - 5$ Step 8: Divide each side by $y - 3$

$x = \dfrac{7y - 5}{y - 3}$

Ⓠ **Let's do some GMAT-type exercises.**

Exercise 1: $2x - 6 = 4$; $x + 3 =$

    A. 5   **D.** 8

    B. 6   **E.** 9

    C. 7

Exercise 2: $x - 9 = 9 - x$; $x =$

    A. 0   **D.** 13.5

    B. 4.5   **E.** 18

    C. 9

Exercise 3: $4x - 17 = 32$; $12x - 51 =$

    A. $12\dfrac{1}{4}$ **D.** 96

    B. $36\dfrac{3}{4}$ **E.** 288

    C. 64

Exercise 4: $\dfrac{xy}{y-x} = 1; x =$

    A. $\dfrac{1}{2}$ D. $\dfrac{y}{y-1}$

    B. 1   E. $\dfrac{y}{1-y}$

    C. $\dfrac{y}{y+1}$

A   **Let's look at the answers.**

Answer 1:   **D:** $x = 5$, but the question asks for $x + 3 = 8$. *Note:* Tests such as the GMAT often ask for $x +$ something instead of just x. Be careful—give what the test wants.

Answer 2:   **C:** $2x = 18$; $x = 9$. The answer is C. It actually can be solved just by looking, since $9 - 9 = 9 - 9$ (or $0 = 0$).

Answer 3:   **D:** We do not solve this equation. We recognize that $12x - 51 = 3(4x - 17) = 3(32) = 96$.

Answer 4:   **C:** By cross-multiplying, we get $xy = y - x$. Then $xy + x = y$, which factors to $x(y + 1) = y$. Therefore, $x = \dfrac{y}{y+1}$.

The GMAT does ask questions like Exercise 4. I know I keep mentioning it, but you must be able to do this without paper and pencil.

LINEAR INEQUALITIES

To review some facts about inequalities:

$a < b$ (read, "a is less than b") means a is to the left of b on the number line.

$a > b$ (read, "a is greater than b") means a is to the right of b on the number line.

$x > y$ is the same as $y < x$.

The notation $x \geq y$ (read, "x is greater than or equal to y") means $x > y$ or $x = y$.

Similarly, $x \leq y$ (read, "x is less than or equal to y"), means $x < y$ or $x = y$.

We solve linear inequalities ($<$, $>$, $\leq$, $\geq$) the same way we solve linear equalities, except when we multiply or divide by a negative, the order reverses.

Example 5: Solve for x: $6x + 2 < 3x + 10$

Solution: $3x < 8$, so $x < \dfrac{8}{3}$. The inequality does not switch because both sides are divided by a positive number (3).

Example 6: Solve for x: $-2(x - 3) \leq 4x - 3 - 7$

Solution: $-2x + 6 \leq 4x - 10$, or $-6x \leq -16$. Thus, $x \geq \dfrac{-16}{-6} = \dfrac{8}{3}$. Here the inequality switches because we divided both sides by a negative number (-6).

Example 7: Solve for x: $8 > \dfrac{x - 2}{-3} \geq 5$

Solution: We multiply through by -3, and both inequalities switch. We get $-24 < x - 2 \leq -15$. If we add 2 to each part to get a value for x alone, the final answer is $-22 < x \leq -13$.

Ⓠ Now, let's do some more exercises.

Exercise 5: If $3x + 4 > 17$, $3x + 7 >$

    A. $\dfrac{13}{3}$ D. 17

    B. $\dfrac{22}{3}$ E. 20

    C. 14

Exercise 6: If $3x + 4y < 5$; $x <$

    A. $5 - 4y - 3$ D. $\dfrac{5 - 4y}{3}$

    B. $\dfrac{5}{4}y - 3$ E. $\dfrac{5}{3} - 4y$

    C. $\dfrac{5}{12}y$

Exercise 7: $x > 0$ and $y > 0$. The number of ordered pairs of whole numbers (x, y) such that $2x + 3y < 9$ is

A. 1   D. 4

B. 2   E. 5

C. 3

 Let's look at the answers.

Answer 5:   **E:** We don't actually have to solve this one. $3x + 7 = (3x + 4) + 3 > 17 + 3$, or 20.

Answer 6:   **D:** $3x + 4y < 5$ is the same as $3x < 5 - 4y$. Dividing by 3, we get $\dfrac{5 - 4y}{3}$.

Answer 7:   **C:** We must substitute numbers. (1, 1) is okay since $2(1) + 3(1) < 9$; (2, 1) is okay since $2(2) + 3(1) < 9$; (1, 2) is okay since $2(1) + 3(2) < 9$; and that's all.

We will do more on ordered pairs later in the book. As we have seen already, some questions overlap more than one topic.

ABSOLUTE VALUE EQUALITIES

Absolute value is the magnitude, without regard to sign. You should know the following facts about absolute value:

$|3| = 3$, $|-7| = 7$, and $|0| = 0$

$|u| = 6$ means that $u = 6$ or -6.

$|u| = 0$ always means $u = 0$.

$|u| = -17$ has no solutions since the absolute value is never negative.

Example 8:   Solve for x: $|2x - 5| = 7$

Solution: Either $2x - 5 = 7$ or $2x - 5 = -7$. So $x = 6$ or $x = -1$.

Note   *This kind of problem always has two answers.*

Example 9:   Solve for x: $|5x + 11| = 0$.

Solution: $5x + 11 = 0$; $x = -\dfrac{11}{5}$.

Note   *This type of problem (absolute value equals 0) always has one answer.*

Those of you with some math background know there is a lot more to absolute value. This section and the next, however, are all you need for the GMAT.

Q   **Let's do a few more exercises.**

Exercise 8: $|2x + 1| = |x + 5|$; $x =$

A. -2 D. 4 and -2

B. 0   E. 0 and 4

C. 4

Exercise 9: $|x - y| = |y - x|$. This statement is true:

A. For no values   D. Only for all integers

B. Only if $x = y = 0$ E. For all real numbers

C. Only if $x = y$

Exercise 10:   If $4|2x + 3| = 11$, $8|2x + 3| + 5 =$

A. $\dfrac{11}{4}$ D. 27

B. $\dfrac{31}{4}$ E. 110

C. 22

A   **Let's look at the answers.**

Answer 8:   D: $2x + 1 = x + 5$ or $2x + 1 = -(x + 5)$, for which the answers are 4 and -2 respectively.

Answer 9:   E: For example, if $x = 3$ and $y = -2$, $|3 - (-2)| = |-2 - 3| = 5$

Answer 10:   D: We don't have to solve this at all. If $4|2x + 3| = 11$, then $8|2x + 3| = 2(11) = 22$. Adding 5, we get 27.

ABSOLUTE VALUE INEQUALITIES

If we talk about integers $|u| \leq 3$, we have $|u| = -3, -2, -1, 0, 1, 2,$ and 3. So $|u| \leq a$ means $-a \leq u \leq a$, where $a > 0$.

Example 10: Solve for x: $|2x - 3| < 11$

Solution: $-11 < 2x - 3 < 11$. Adding 3 to each piece means $-8 < 2x < 14$. Dividing by 2 gives $-4 < x < 7$.

If we talk about integers $|u| \geq 4$, we have $u = 4, 5, 6, \ldots$ and $-4, -5, -6, \ldots$ So $|u| \geq a$ means $u \geq a$ or $u \leq -a, a > 0$.

Example 11: Solve for x: $|x - 7| > 4$.

Solution: $x - 7 > 4$ or $x - 7 < -4$. The two parts of the answer are $x > 11$ or $x < 3$.

Q **Let's do some more exercises:**

Exercise 11: If $5 \leq |x| \leq 5, x =$

    A. 0   D. no values

    B. 5   E. all values

    C. -5 and 5

Exercise 12: $|x - 5| \geq -5$ if $x =$

    A. 0   D. no values

    B. 5   E. all values

    C. -5 and 5

Exercise 13: $|x + 5| \leq -5$ if $x =$

    A. 0   D. no values

    B. 5   E. all values

    C. -5 and 5

Let's look at the answers.

Answer 11:   **C:** Both $|-5|$ and $|5|$ equal 5.

Answer 12:   **E:** The absolute value is always greater than any negative number because it is always greater or equal to zero.

Answer 13:   **D:** The absolute value can't be less than a negative number.

QUADRATIC EQUATIONS

Quadratic equations, equations involving the square of the variable, can be solved in three principal ways: factoring, taking the square root, and using the quadratic formula. Another name for the solution of any equation is a **root**.

Solving Quadratics by Factoring

Solving quadratic equations by factoring is based on the fact that if $a \times b = 0$, then either $a = 0$ or $b = 0$.

Example 12:   Solve for all values of x: $x(x - 3)(x + 7)(2x + 1)(3x - 5)(ax + b)(cx - d) = 0$.

Solution: Setting each factor equal to 0 (better if you can do it just by looking), we get $x = 0, 3, -7, -\dfrac{1}{2}, \dfrac{5}{3}, -\dfrac{b}{a}$, and $\dfrac{d}{c}$.

Solving Quadratics by Taking the Square Root

If the equation is of the form $x^2 = c$, with no x term, we just take the square root: $x = \pm\sqrt{c}$.

   Remember that $\sqrt{9} = 3$, $-\sqrt{9} = -3$, $\sqrt{-9}$ is not real, and if $x^2 = 9$, then $x = \pm 3$!

Example 13:   Solve for all values of x:

a. $x^2 - 7 = 0$

b. $ax^2 - b = c$, where $a, b, c > 0$.

Solutions: a. $x^2 = 7$; $x = \pm\sqrt{7}$

b. $ax^2 = b + c$; $x^2 = \dfrac{b + c}{a}$, so $x = \pm\sqrt{\dfrac{b + a}{c}}$, or $\pm\dfrac{\sqrt{c(b + a)}}{c}$

Solving Quadratics by Using the Quadratic Formula

The quadratic formula states that if $ax^2 + bx + c = 0$, then

$$x = \frac{-b \pm \sqrt{b^2 - 4ac}}{2a},$$

where a is the coefficient of the x^2 term, b is the coefficient of the x term, and c is the number term.

Example 14: Solve $3x^2 - 5x + 2 = 0$ by using the quadratic formula.

Solution: $a = 3, b = -5, c = 2.$ $x = \dfrac{-(-5) \pm \sqrt{(-5)^2 - 4(3)(2)}}{2(3)} = \dfrac{5 \pm 1}{6}. x_1$ (read as

"x sub one," the first answer) $= \dfrac{5 + 1}{6} = 1; x_2$ (read, "x sub two," the second

answer) $= \dfrac{5 - 1}{6} = \dfrac{2}{3}.$

Example 15: Solve $3x^2 - 5x + 2 = 0$ by factoring.

Solution: This is the same problem as Example 14. $3x^2 - 5x + 2 = (3x - 2)(x - 1)$
$= 0$, so $x = 1, \dfrac{2}{3}.$

Factoring is preferred; using the quadratic formula takes too long.

Before we go to the next set of exercises, I suppose most of you know the quadratic formula, but few have seen it shown to be true. The teacher in me has to show you, even if you don't care.

$ax^2 + bx + c = 0$ — The coefficient of x^2 must be 1.

$x^2 + \dfrac{b}{a}x = -\dfrac{c}{a}$ — Complete the square; this means taking half the coefficient of x, squaring it, adding it to both sides.

$x^2 + \dfrac{b}{a}x + \left(\dfrac{b}{2a}\right)^2 = \left(\dfrac{b}{2a}\right)^2 - \dfrac{c}{a}$ — Factor the left side and do the algebra on the right side.

$\left(x + \dfrac{b}{2a}\right)^2 = \dfrac{b^2}{4a^2} - \dfrac{c(4a)}{a(4a)} = \dfrac{b^2 - 4ac}{4a^2}$ — Multiply the last term by $\dfrac{4a}{4a}$ and combine terms.

$\left(x + \dfrac{b}{2a}\right)^2 = \dfrac{b^2}{4a^2} - \dfrac{4ac}{4a^2} = \dfrac{b^2 - 4ac}{4a^2}$ — Take the square root of both sides.

$x + \dfrac{b}{2a} = \dfrac{\pm\sqrt{b^2 - 4ac}}{2a}$ — Solve for x and simplify.

$x = -\dfrac{b}{2a} \dfrac{\pm\sqrt{b^2 - 4ac}}{2a} = \dfrac{-b \pm \sqrt{b^2 - 4ac}}{2a}$

The quadratic formula is really true! You should have questioned it in high school. Hopefully, college has taught you to question everything.

The quadratic equation has been included here for completeness. Quadratics on the GMAT can usually be solved by factoring or taking the square root.

Let's go on.

CHAPTER 7: *Word Problems in One Unknown*

"It is necessary to study the words of math. Only then can you truly understand all."

I consider this section the most important section of the book. Although the book is filled with skills that may be on the GMAT, most of the questions will not be pure math questions, but will have words that you must interpret and then do some math. Unfortunately, most high schools have de-emphasized these kinds of problems. In this chapter, I will try to make these dreaded "word problems" easy to understand.

First, we'll look at the words you'll need to know. Next, we'll go over the more likely problems to show up on the GMAT. We'll then go over the other types of word problems. Finally, we'll review some common measurements.

BASICS

As we know, the answer in **addition** is the **sum**. Other words that indicate addition are **plus**, **more**, **more than**, **increase**, and **increased by**. You can write all sums in any order since addition is commutative.

The answer in **multiplication** is the **product**. Another word that is used is **times**. Sometimes the word "**of**" indicates multiplication, as we shall see shortly. **Double** means to multiply by two, and **triple** means to multiply by three. Since multiplication is also commutative, we can write any product in any order.

Division's answer is called the **quotient**. Another phrase that is used is **divided by**.

The answer in **subtraction** is called the **difference**. Subtraction can present a reading problem because $4 - 6 \neq 6 - 4$, so we must be careful to subtract in the correct order. Example 1 shows how some subtraction phrases are translated into algebraic expressions.

Example 1:	Phrases	Expressions
a.	The difference between 9 and 5	$9 - 5$
	The difference between m and n	$m - n$
b.	Five minus two	$5 - 2$
	m minus n	$m - n$
c.	Seven decreased by three	$7 - 3$
	m decreased by n	$m - n$
d.	Nine diminished by four	$9 - 4$
	m diminished by n	$m - n$
e.	Ten less two	$10 - 2$
	m less n	$m - n$
f.	Ten less than two;	$2 - 10$
	m less than n	$n - m$
g.	Three from five	$5 - 3$
	m from n	$n - m$

 One word can make a big difference. In parts e and f of Example 1, m less n means $m - n$; and m less than n means $n - m$. In addition, m is less than n means $m < n$. You must read carefully!

The following words usually indicate an equal sign: is, am, are, was, were, the same as, equal to.

You also must know the following phrases for inequalities: at least ($\geq$), not more than ($\leq$), over ($>$), and under ($<$).

Example 2:   Write the following in symbols:

Problem	Solution
a. m times the sum of q and r	$m(q + r)$
b. Six less the product of x and y	$6 - xy$
c. The difference between c and d divided by f	$\dfrac{c - d}{f}$
d. b less than the quotient of r divided by s	$\dfrac{r}{s} - b$
e. The sum of d and g is the same as the product of h and r	$d + g = hr$
f. x is at least y	$x \geq y$
g. Zeb's age n is not more than 21	$n \leq 21$
h. I am over 30 years old, where "l" represents "my age."	$l > 30$
i. Most people are under seven feet tall, where "p" represents "most people"	$p < 7$

RATIOS

Comparing two numbers is called a **ratio**. The ratio of 3 to 5 is written two ways: $\dfrac{3}{5}$ or $3 : 5$ (read, "the ratio of 3 to 5").

Example 3:   Find the ratio of 5 ounces to 2 pounds.

Solution: The ratio is $\dfrac{5}{32}$, since 16 ounces are in a pound.

Example 4:   A board is cut into two pieces that are in the ratio of 3 to 4. If the board is 56 inches long, how long is the longer piece?

Solution: If the pieces are in the ratio $3 : 4$, we let one piece equal $3x$ and the other $4x$. The equation, then, is $3x + 4x = 56$; so $x = 8$; and the longer piece is $4x = 32$ inches.

I have asked this problem many, many times. Almost no one has ever gotten it correct—not because it is difficult, but because no one does problems like this anymore.

CONSECUTIVE INTEGERS

If there are any "fun" word problems, they would have to do with consecutive integers. Let's recall the following facts about integers, most of which you know without having to even think about them: Integers: $-3, -2, -1, 0, 1, 2, 3, 4, \ldots$

Evens: $-6, -4, -2, 0, 2, 4, 6, 8, \ldots$.

Odds: $-5, -3, -1, 1, 3, 5, 7, \ldots$

If we let $x =$ integer, then $x + 1$ represents the next consecutive integer, and $x + 2$ represents the next consecutive integer after that. Then $x + (x + 1) + (x + 2) = 3x + 3$ is the sum of three consecutive integers, where x is the smallest and $x + 2$ is the largest in the group.

If $y =$ an even integer, then $y + 2$ is the next consecutive even integer; $y + 4$ and $y + 6$ are the next consecutive integers after that. Then the sum of four consecutive even integers is $y + (y + 2) + (y + 4) + (y + 6) = 4y + 12$.

Similarly, if z is an odd integer, the next three odd integers are $z + 2$, $z + 4$, and $z + 6$. This is the same as for even integers, except we start out letting z be odd instead of even.

Example 5:   The sum of three consecutive integers is twice the smallest. What is the smallest integer?

Solution: We have $x + (x + 1) + (x + 2) = 2x$, which simplifies to $3x + 3 = 2x$. So $x = -3$. The integers are -3, -2, and -1; and the smallest is -3.

Note   *Integers can be negative!*

Most consecutive integer problems are done by using tricks, as the next few examples show.

Example 6:   The sum of five consecutive even integers is 210. What is the sum of the smallest two?

Solution: If we have an odd number of consecutive, consecutive even, or consecutive odd integers, the middle number is the average (the mean); So the middle number is given by $\dfrac{210}{5} = 42$. Once we know that, count backwards and forwards to get the others. The five numbers are 38, 40, 42, 44, and 46. The sum of the two smallest is $38 + 40 = 78$.

Example 7: The sum of four consecutive integers is -50. What is the sum of the two largest?

Solution: Dividing -50 by 4, we get -12.5. The four consecutive integers are the closest integers to -12.5, namely -14, -13, -12, and -11. The sum of the two largest, -12 and -11, is -23.

AGE

Similar to consecutive integer problems are age problems. We just have to think about it logically.

Example 8: p years ago, Mary was q years old; in r years she will be how many years old?

Solution: The secret is age now. If Mary was q years old p years ago, now she is $p + q$. Then r years in the future, she will be $p + q + r$. If necessary, substitute numbers for p, q, and r to see how this works.

SPEED

We are familiar with speed being given in miles per hour (mph), so it is easy to remember that $\text{speed} = \dfrac{\text{distance}}{\text{time}}$, or $r = \dfrac{d}{t}$, where r stands for rate (the speed). Use this relationship, or the equivalent ones: $d = rt$ or $t = \dfrac{d}{r}$, to do word problems involving speed.

Example 9: Sue drives for 2 hours at 60 mph and 3 hours at 70 mph. What is her average speed?

Solution: Sue's average speed for the whole trip is given by $r = \dfrac{d}{t}$, where d is the total distance and t is the total time. Note that her average speed is *not* the average of the speeds. Use $d = rt$ for each part of her trip to get the total distance. The total distance is $60(2) + 70(3) = 330$ miles. The total time is 5 hours. So Sue's average speed is $r = \dfrac{330}{5}$ mph $= 66$ mph.

Example 10:   Don goes 40 mph in one direction and returns at 60 mph. What is his average speed?

Solution: Notice that the problem doesn't tell the distance. It doesn't have to; the distance in each direction is the same, since it is a round trip. We can take any distance, so let's choose 120 miles, the LCM of 40 and 60. Then the time going is $\frac{120}{40} = 3$ hours, and the time returning is $\frac{120}{60} = 2$ hours. The average speed is the total distance divided by the total time, $\frac{2(120)}{3 + 2} = \frac{240}{5} = 48$ mph.

We actually don't have to choose a number for the distance, however. We could use x. Just for learning's sake, we will do this same problem (Example 10) using x as the distance. The time going is $\frac{x}{40}$, and the time returning is $\frac{x}{60}$. Then we have:

$$\text{Total speed} = \frac{\text{total distance}}{\text{total time}} = \frac{2x}{\frac{x}{40} + \frac{x}{60}} = \frac{120(2x)}{120(\frac{x}{40} + \frac{x}{60})} = \frac{240x}{5x} = 48 \text{ mph}$$

Example 11:   A plane leaves Indianapolis traveling west. At the same time, a plane traveling 30 mph faster leaves Indianapolis going east. After two hours the planes are 2,000 miles apart. What is the speed of the faster plane?

Solution: A chart and a picture are best for problems like this, but you must be able to picture this in your mind.

	r	t	d
W	x	2	$2x$
E	$x + 30$	2	$2(x + 30)$

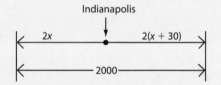

We let x = the speed of the plane going west; then $x + 30$ is the speed of the eastern-going plane. The time for each is 2 hours. Since $rt = d$, the distances are as shown in the above chart. According to the picture, $2x + 2(x + 30) = 2000$. So $x = 485$; $x + 30 = 515$ mph.

The problem is the same if the planes are starting at the ends and flying toward each other.

Let's do some basic exercises.

Exercise 1: Four more than a number is seven less than triple the number. The number is

A. 4

B. 5.5

C. 7

D. 9

E. 11

Note *The word "number" does not necessarily mean an integer or even necessarily a positive number.*

Exercise 2: Mike must have at least an 80 average but less than a 90 average to get a *B*. If he received scores of 98, 92, and 75 on the first three tests, which of these grades on the fourth test will give him a *B*?

A. 42

B. 54

C. 66

D. 98

E. 100

Exercise 3: Nine less than a number is the same as the difference between nine and the number. The number is

A. 18

B. 9

C. 0

D. −9

E. All numbers are correct.

Exercise 4: Seven consecutive odd numbers total −77. The sum of the largest three is

A. −21

B. −27

C. −33

D. −39

E. −45

Exercise 5: For three consecutive integers, the sum of the squares of the first two equals the square of the largest. There are two sets of answers. The sum of all six integers is

A. 0

B. 3

C. 6

D. 12

E. 24

Exercise 6: Ed goes 20 mph in one direction and 50 mph on the return trip. His average speed is

A. 25 mph

D. 30 mph

B. $27\frac{2}{7}$ mph

E. $30\frac{6}{7}$ mph

C. $28\frac{4}{7}$ mph

Exercise 7: The angles of a triangle are in the ratio of $3:5:7$. The largest angle is

A. $12°$

D. $84°$

B. $36°$

E. $108°$

C. $60°$

Exercise 8: b years in the future, I will be c years old. How old was I six years in the past?

A. $b - c - 6$

D. $b - c + 6$

B. $c - b - 6$

E. $b + c - 6$

C. $c - b + 6$

Exercise 9: The value of d dimes and q quarters in pennies is

A. $d + q$

D. $10d + 25q$

B. dq

E. $35dq$

C. $250dq$

Exercise 10: A car leaves Chicago at 2 p.m. going west. A second car leaves Chicago at 5 p.m., going west at 30 mph faster. At 7 p.m., the faster car hits the slower one. The accident occurred after how many miles?

A. 200

D. 40

B. 100

E. 20

C. 60

Exercise 11: Meg is six times as old as Peg. In 15 years, Meg will be three times as old as Peg. Meg's age now is

A. 10

D. 75

B. 25

E. 90

C. 60

Exercise 12: A fraction, when reduced, is $\frac{2}{3}$. If 6 is added to the numerator and 14 is added to the denominator, the fraction reduces to $\frac{3}{5}$. The sum of the original numerator and denominator is

    A. 60   **D.** 90

    B. 70   **E.** 100

    C. 80

 Let's look at the answers.

Answer 1: **B:** Let's break this one down into small pieces. Four more than a number is written as $n + 4$ (or $4 + n$). Seven less than triple the number is $3n - 7$ (the only correct way). "Is" means equals, so the equation is $n + 4 = 3n - 7$. Solving, we get $n = 5.5$.

Answer 2: **C:** The "setup" to do this problem is $80 \leq \dfrac{(98 + 92 + 75 + x)}{4} < 90$.

However, $80(4) = 320$ total points for a minimum, and it must be less than $90(4) = 360$ points. So far, Mike has $98 + 92 + 75 = 265$ points; $265 + 66 = 321$ points. (However, choices D or E will result in a grade of A, and I'm sure Mike wouldn't object to that.)

Answer 3: **B:** $x - 9 = 9 - x$; $x = 9$.

Answer 4: **A:** The middle one is $-\dfrac{77}{7} = -11$. The three largest ones are thus -9, $-7, -5$, and their sum is -21.

Answer 5: **D:** $x^2 + (x + 1)^2 = (x + 2)^2$, which simplifies to $x^2 - 2x - 3 = 0$; $(x - 3)(x + 1) = 0$. The solution set is $x = 3$ or $x = -1$. For $x = 3$, the integers are 3, 4, 5; for $x = -1$, the integers are $-1, 0, 1$. The sum of all six is $3 + 4 + 5 + (-1) + 0 + 1 = 12$.

Answer 6: **C:** If we assume a 100-mile distance, the original trip was 5 hours, and the return trip was 2 hours. $r = \dfrac{d}{t} = \dfrac{200}{7} = 28\dfrac{4}{7}$ mph.

Answer 7: D: $3x + 5x + 7x = 180$, so $x = 12°$. The largest angle is $7x = 84°$.

Answer 8: B: My age now is $c - b$, so six years ago it was $(c - b) - 6$.

Answer 9: D: Dimes are 10 cents each, so the value is $10d$, where d is the number of dimes. Similarly, the value of quarters is $25q$, where q is the number of quarters.

Answer 10: B: Let's construct a chart and picture.

	r	t	d
Slower	x	5	5x
Faster	x + 30	2	2(x + 30)

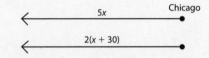

The rate of the slower car is x, and the time of the slower car is 7 p.m. -2 p.m., or 5 hours. The rate of the faster car is $x + 30$; and its time is 7 p.m. $-$ 5 p.m., or 2 hours. When they crashed, their distances were equal, so $5x = 2(x + 30)$. Then $x = 20$; and the total distance is 5(20) or $2(20 + 30) = 100$ miles.

Answer 11: C:

	Age now	Age in 15 years
Peg	x	x + 15
Meg	6x	6x + 15

We let $x = $ Peg's (the younger one's) age. Meg's age is thus $6x$. In 15 years, Meg's age ($6x + 15$) will be (equals) three times Peg's age ($3(x+15)$). So the equation is $6x + 15 = 3x + 45$, and $x = 10$. Meg's age now is $6x = 60$.

Answer 12: A: The fraction can be written as $\dfrac{2x}{3x}$. So $\dfrac{2x + 6}{3x + 14} = \dfrac{3}{5}$. Cross-multiplying, we get $5(2x + 6) = 3(3x + 14)$, so $x = 12$; the original numerator is $2x = 24$ and the original denominator is $3x = 36$, and their sum is $24 + 36 = 60$.

WORK

The basic idea of work problems is that if a job can be done in x hours, then each hour, the amount that is done is $\frac{1}{x}$ of the job. In 4 hours, for example, a job that takes 6 hours to do is $\frac{4}{6} = \frac{2}{3}$ done. The whole job done is represented by the number 1.

Example 12: Rob can do a job in 8 hours and Nan can do the same job in 4 hours. Together, they can do the job in how many hours?

Solution: I always say there are two answers to this problem. The first is that they start watching TV and the job never gets done. However, this is how to do the real problem. If a job can be done in 8 hours, then the part done in one hour is $\frac{1}{8}$; in three hours, it is $\frac{3}{8}$; and in x hours, it is $\frac{x}{8}$. The part done in x hours by Rob is $\frac{x}{8}$, and similarly, the part done by Nan in x hours is $\frac{x}{4}$. The part done by Rob plus the part done by Nan is the whole job, so $\frac{x}{8} + \frac{x}{4} = 1$. Multiplying by 8, we get $x + 2x = 8$. Thus, $x = 2\frac{2}{3}$. It would take them $2\frac{2}{3}$ hours to do the job together.

Example 13: Sandy takes twice as long to do a job as Randy. They finish the job together in 3 hours. How long would it take Randy to do the job alone?

Solution: The number of hours it takes Randy is x hours, so he does $\frac{3}{x}$ of the job in 3 hours. Sandy takes twice as long ($2x$ hours) to do the job, so she does $\frac{3}{2x}$ of the job in 3 hours. Therefore, $\frac{3}{x} + \frac{3}{2x} = 1$ is the equation representing the work done in 3 hours. Multiplying by $2x$, we get $6 + 3 = 2x$, or $x = 4.5$. It would take Randy 4.5 hours to do the job alone.

MIXTURES

Mixture problems are easier done with charts.

Example 14: Walnuts selling at $6.00 a pound are mixed with 24 pounds of almonds at $9.00 a pound to give a mixture selling at $7.00 a pound. How many pounds of walnuts are used?

Solution:

	Cost per Pound	× Number of Pounds	= Total Cost
Walnuts	6	x	$6x$
Almonds	9	24	216
Mixture	7	$x + 24$	$7(x + 24)$

We set up a chart for the cost. The columns are cost per pound, the number of pounds, and the total cost. We let x equal the number of pounds of walnuts. The total pounds of walnuts plus the total pounds of almonds is the total weight of the mixture. The equation is the cost of the walnuts plus the cost of the almonds equals the cost of the mixture: $6x + 216 = 7(x + 24)$, so $x = 48$ pounds of walnuts.

However, since you cannot make a chart on the GMAT, you might be able to keep a chart in your head and use logic to solve this problem. Looking at the 24 pounds of almonds at $9.00, the mixture is $7.00, "twice as close to $6.00 as to $9.00." So we need twice as many pounds of walnuts, 48 pounds. If the mixture was $8.00, we would need 12 pounds of walnuts.

Coin problems are just like mixture problems, except we are working with money, not nuts and bolts (or walnuts and almonds).

Example 15: There are 40 coins in nickels and dimes totaling $2.80. How many nickels are there?

Solution:

	Value per Coin	× Number of Coins	= Total Value
Nickels	5	x	$5x$
Dimes	10	$40 - x$	$10(40 - x)$
Mixture	—	40	280

The problem is done in pennies. The total number of coins is 40. If there are x nickels, there are $40 - x$ dimes. The value of x nickels is $5x$. The value of $40 - x$ dimes is $10(40 - x)$. The values of nickels plus dimes is the total value: $5x + 400 - 10x = 280$, so $x = 24$ nickels.

The problem also could be done by trial and error. Twenty of each coin would mean $3.00, so we need more nickels. It must be an even number since the total is $2.80. If we first try 22 and then 24, we would get the correct answer.

Ⓠ Let's do some more exercises.

Exercise 13: Water is poured into a tank at the same time a pipe is opened that drains the tank. If the tank is filled in 10 hours and the tank can empty in 15 hours, and the tank starts empty, how many hours does it take to fill the tank?

A. 20   D. 60

B. 30   E. 120

C. 40

Exercise 14: Adult tickets cost $10 and children's tickets cost $5. If 100 tickets are sold and $800 is taken in, how many adult tickets are sold?

A. 50   D. 75

B. 60   E. 80

C. 70

Exercise 15: Sid is twice as old as Rex. Ten years ago, Sid was four times as old as Rex. How old is Sid today?

A. 5   D. 30

B. 10   E. 60

C. 20

Exercise 16: The number of quarters is four more than twice the number of dimes. If the total is $7.00, how many dimes are there?

A. 10   D. 24

B. 16   E. 34

C. 20

Exercise 17: Fred leaves Fort Worth by car traveling north. Two hours later, Jim also leaves Fort Worth going north, but 20 mph slower. After six more hours, they are 300 miles apart. Fred's speed is

A. 60 mph   D. 90 mph

B. 70 mph   E. 100 mph

C. 80 mph

Exercise 18: Deb can do a job alone in 6 hours. After working alone for two hours, she is joined by Sue who can do the job alone in 8 hours. They work together and finish the job. How many hours did Deb work?

A. $3\frac{3}{7}$ D. $4\frac{1}{2}$

B. 4   E. 5

C. $4\frac{2}{7}$

 Let's look at the answers.

Answer 13: B: $\frac{x}{10} - \frac{x}{15} = 1$. We use a minus sign because it empties. Solving, we get $x = 30$.

Answer 14:

	Value per Tickets	× Number of Tickets =	Total Value
Adult	10	x	10x
Child	5	100 − x	5(100 − x)
Mixture	—	100	800

As the chart indicates, this is similar to the mixture problem, Example 15 in this chapter. The equation is $10x + 500 - 5x = 800$, so $x = 60$.

Answer 15:

	Age now	Age 10 years ago
Sid	$2x$	$2x - 10$
Rex	x	$x - 10$

D: According to the chart and the problem, $2x - 10 = 4(x - 10)$, so $x = 15$. Sid's age is $2x = 30$.

Answer 16: **A:** A chart is probably unnecessary. We have x dimes and $(2x + 4)$ quarters. The equation is $10x + 25(2x + 4) = 700$, so $x = 10$.

Answer 17:

	r	$\times$ t	$=$ d
Fred	x	8	$8x$
Jim	$x - 20$	6	$6(x - 20)$

Fred goes at x mph for 8 hours. Jim goes at $x - 20$ mph for 6 hours. Since they are going in the same direction, we get $8x - 6(x - 20) = 300$, or $x = 90$. Notice that 300 is the *difference* in their distances, not the distance they traveled, so we use subtraction.

Answer 18: **C:** Deb can do the job in 6 hours; her part is $\dfrac{x}{6}$. Sue worked two hours less and can do the job in 8 hours; her part is $\dfrac{x - 2}{8}$. The equation is $\dfrac{x}{6} + \dfrac{x - 2}{8} = 1$, so $x = 2\dfrac{2}{7}$. Deb spent $2\dfrac{2}{7}$ hours working with Sue, plus the 2 hours she worked alone, or $4\dfrac{2}{7}$ hours total.

You must try to do these problems without paper and pencil. That requires reviewing them over and over.

MEASUREMENTS

You might want to review some basic measurements and how to convert a few.

Linear: 12 inches = 1 foot; 3 feet = 1 yard; 5,280 feet = 1 mile.

Liquid: 8 ounces = 1 cup; 2 cups = 1 pint; 2 pints = 1 quart; 4 quarts = 1 gallon

Weight: 16 ounces = 1 pound; 2,000 pounds = 1 ton.

Dry measure: 2 pints = 1 quart; 8 quarts = 1 peck; 4 pecks = 1 bushel. If I love you a bushel and a peck, it would be 5 pecks or 40 dry quarts.

Metric: 1,000 grams in a kilogram; 1,000 liters in a kiloliter; 1,000 meters in a kilometer;

1,000 millimeters = 1 meter; 100 centimeters = 1 meter; 10 millimeters = 1 centimeter

When doing conversions, we pay particular attention to the units, canceling them when doing the multiplications. A good guide is: to go from large to small, multiply; from small to large, divide (or multiply by the reciprocal).

Example 16: Change 30 kilograms 20 grams to milligrams.

Solution: This is going from large to small, so multiply by the conversions:

$$\frac{30 \text{ kg}}{1} \times \frac{1000 \text{ g}}{1 \text{ kg}} \times \frac{1000 \text{ mg}}{1 \text{ g}} + \frac{20 \text{ g}}{1} \times \frac{1000 \text{ mg}}{1 \text{ g}} = 30,020,000 \text{ mg}$$

 Notice how the measurements (g, kg) cancel.

Example 17: Change 90 miles per hour into feet per second.

Solution: Now we are going from small to large, so multiply by the reciprocals

(same as dividing); to change from miles to feet, though, we are going

from large to small, so just multiply by $\dfrac{5,280 \text{ feet}}{1 \text{ mile}}$:

$$\frac{90 \text{ miles}}{\text{hour}} \times \frac{1 \text{ hour}}{60 \text{ minutes}} \times \frac{1 \text{ minute}}{60 \text{ seconds}} \times \frac{5280 \text{ feet}}{1 \text{ mile}} = \frac{132 \text{ feet}}{\text{sec}}$$

 Each fraction after the first is equivalent to 1. When we multiply by 1, the value doesn't change. Again, the measurements cancel, and we wind up with feet per second.

INTEREST

Last, we need to talk a little about simple interest. We know that the interest is equal to the principal times (annual) rate times time (in years). In symbols $i = prt$. The total amount of money is the principal plus the interest, or $A = p + i = p + prt = p(1 + rt)$.

Example 18: Suppose we invest $20,000 at simple interest at 12% for 3 months. How much money do we have?

Solution: $A = p + prt = 20,000 + 20,000(.12)\left(\dfrac{1}{4}\right) = \$20,600$. Of course, we wouldn't normally invest at simple interest. However, a safe 12% interest would be great.

Example 19: Suppose we invest $10,000 at 10% for a year, compounded every six months. How much do we have after a year?

Solution: For the first six months, or half a year, $A = p + prt = 10,000 + 10,000(.10)\left(\dfrac{1}{2}\right) = \$10,500$. After the second six months, $A = p + prt = 10,500 + 10,500(.10)\left(\dfrac{1}{2}\right) = \$11,025$, the amount after one year. This is $25 more than if we had simple interest for the full year. That is why compounding continuously is most desirable—we literally earn interest on our interest.

CHAPTER 8: *Working with Two or More Unknowns*

"*Understanding more complex problems will be gratifying to you.*"

SOLVING SIMULTANEOUS EQUATIONS

We can solve two equations in two unknowns, also known as **simultaneous equations**, in five basic ways. Only two are practical for this exam: **substitution** and **elimination**. Sometimes, we use a combination of these two.

Substitution

In substitution, we find an unknown with a coefficient of 1, solve for that variable, and substitute it in the other equation.

Example 1: Solve for x and y:

$$3x + 4y = 4 \qquad (1)$$
$$x - 5y = 14 \qquad (2)$$

Solution: In equation (2) $x = 5y + 14$. Substituting this into equation (1), we get $3(5y + 14) + 4y = 4$. Solving, we get $y = -2$; $x = 5y + 14 = 5(-2) + 14 = 4$. The answer is $x = 4$ and $y = -2$.

We can check by substituting these values into the original equations to see that they are solutions to both equations.

If the coefficient of none of the terms is 1, the elimination method is better.

Elimination

There are several ways to eliminate one of the variables, as seen in the following examples. Once we have eliminated one of the variables, we can solve for the other variable by using substitution.

Example 2: $2x + 3y = 12$

$5x - 3y = 9$

Solution: If we add the equations, term by term, we can eliminate the y term. We get $7x = 21$, so $x = 3$; substituting $x = 3$ into either equation, we get $y = 2$. So the answer is $x = 3$, $y = 2$.

If addition doesn't work, try subtraction.

Example 3: $5x + 4y = 14$

$5x - 2y = 8$

Solution: By subtracting, we get $6y = 6$; so $y = 1$; then by substituting, we get $x = 2$.

If adding or subtracting doesn't work, we must find two numbers that when we multiply the first equation by one of them and the second equation by the other, and then add (or subtract) the resulting equations, one letter is eliminated.

Example 4: $5x + 3y = 11$ (1)

$4x - 2y = 22$ (2)

Solution: To eliminate x, multiply the first equation by 4 and the second by -5; then add.

$$4(5x + 3y) = 4(11) \quad \text{or} \quad 20x + 12y = 44$$

$$-5(4x - 2y) = -5(22) \quad \text{or} \quad -20x + 10y = -110$$

Adding, we get $22y = -66$; so $y = -3$. We could substitute now, but we could also eliminate y by multiplying the original equation (1) by 2 and equation (2) by 3. Let's do that.

$$2(5x + 3y) = 2(11) \quad \text{or} \quad 10x + 6y = 22$$

$$3(4x - 2y) = 3(22) \quad \text{or} \quad 12x - 6y = 66$$

Adding, we get $22x = 88$; so $x = 4$. The answer is $x = 4$, $y = -3$.

Practice in Solving Simultaneous Equations

For those of you who are curious, the other three basic ways of solving simultaneous equations are by using graphs, matrices, or determinants.

Ⓠ **Let's do some exercises.**

Exercise 1: If $x = y + 3$ and $y = z + 7$, x (in terms of z) =

A. $z - 10$

B. $z - 4$

C. z

D. $z + 4$

E. $z + 10$

Exercise 2: Two apples and 3 pears cost 65 cents, and 5 apples and 4 pears cost $1.10. Find the cost in cents of one pear:

A. 10

B. 15

C. 20

D. 25

E. 30

Exercise 3: As in Exercise 2, 2 apples and 3 pears cost 65 cents, and 5 apples and 4 pears cost $1.10. Find the cost of one pear and one apple together:

A. 10

B. 15

C. 20

D. 25

E. 30

Exercise 4: Find $x + y$:

$7x + 4y = 27$

$x - 2y = -3$

A. 1

B. 3

C. 5

D. 7

E. 9

Exercise 5: For lunch, Ed buys 3 hamburgers and one soda for $12.50, and Mei buys one hamburger and one soda for $5.60. How much does Ed pay for his hamburgers?

A. $2.15

B. $3.45

C. $6.90

D. $10.35

E. $18.10

Let's look at the answers.

Answer 1:   E: $x = y + 3 = (z + 7) + 3 = z + 10$.

Answer 2:   B:

$2a + 3p = 65$

$5a + 4p = 110$

In solving for p, eliminate a by multiplying the top equation by 5 and the bottom by -2.

$5(2a + 3p) = 5(65)$ or $10a + 15p = 325$

$-2(5a + 4p) = -2(110)$ or $-10a - 8p = -220$

Adding, we get $7p = 105$; $p = 15$.

Answer 3:   D: Much more often, we get a problem like this. The equations are the same as in Exercise 2, but rather than asking for the cost of one apple or the cost of one pear, this exercise asks for the cost of one apple plus one pear. The trick is simply to add the original equations. We then get $7a + 7p = 175$. Dividing both sides by 7, we get $a + p = 25$.

Answer 4:   C: Less frequently, when adding doesn't work, try subtracting. If we subtract, the difference becomes $6x + 6y = 30$. So $x + y = 5$.

Answer 5:   D: The equations are

$3h + s = 12.50$

$h + s = 5.60$

Subtracting, we get $2h = 6.90$, so $h = 3.45$. Ed's three hamburgers cost $10.35.

NEW WORD PROBLEMS

Let's try some "new" word problems. The only true difference from what we saw in Chapter 7 involves tens and units digit problems.

Digit Problems

If we have a two-digit number, it is represented by t for the tens digit and u for the units digit. For example, for 68, $t = 6$ and $u = 8$. The value of the number is represented by $10t + u$. The number 68 would thus be $10(6) + 8$. The number with the digits reversed is $10u + t$, or $86 = 10(8) + 6$. The sum of the digits is $t + u$, or $6 + 8 = 14$ in this case.

Example 5:   In a two-digit number, the sum of the digits is 8. If the digits are reversed, the new number is 36 more than the original number. What is the number?

Solution: There are two methods to answer this question:

Method A: $t + u = 8$ is the first equation. The new number, $10u + t$ is $(=)$ 36 more than the original number, or $10t + u + 36$. So the equation is $10u + t = 10t + u + 36$, or $9u - 9t = 36$. In this type of problem only, we can divide the equation by 9 and get $u - t = 4$. Since $t + u = 8$, adding these two equations gives $2u = 12$, or $u = 6$, so $t = 2$, and the number is 26.

Method B: Since $t + u = 8$, the only possible answers could be 17, 26, or 35, since the number with the digits reversed is bigger. $71 - 17 \neq 38$, but $62 - 26 = 36$; so the number is 26.

Mixture Problems

Mixture problems with two unknowns are treated similarly to how we did mixture problems with one unknown in the last chapter. It often helps to construct a chart, even though you can do that only mentally on the GMAT.

Example 6:   How many pounds of peanuts at $3.00 a pound must be mixed with $7.00-per-pound cashews to give 20 pounds of a $6.00 mixture?

Solution: Construct a chart with the given information.

	Cost per Pound	Number of Pounds	Total Cost
Peanuts	3	x	$3x$
Cashews	7	y	$7y$
Mixture	6	20	120

Let x = pounds of peanuts and y = pounds of cashews. According to the chart,

$$x + y = 20$$
$$3x + 7y = 120$$

Since we want x, multiply the top equation by -7, and add the result to the second equation.

$$-7x + -7y = -140$$
$$3x + 7y = 120$$

Adding, we get $-4x = -20$, so $x = 5$ pounds of peanuts.

Age Problems

Age problems with two unknowns are treated similarly to how we did age problems with one unknown in the last chapter. It often helps to construct a chart, but again, you must do this mentally on the GMAT.

Example 7:   Joan is 4 times the age of Ben. In 4 years, Joan will be $2\frac{1}{2}$ times the age of Ben. Ben will be how many years old then?

Solution: Construct a chart with the given information.

	Age now	Age in 4 years
Joan	x	$x + 4$
Ben	y	$y + 4$

Let x = Joan's age and y = Ben's age. The first equation is $x = 4y$. In 4 years, the equation is

$$x + 4 = \left(\frac{5}{2}\right)(y + 4), \text{ or } 2(x + 4) = 5(y + 4).$$

By the Distributive Law (Chapter 5),

$$2x + 8 = 5y + 20, \text{ or } 2x - 5y = 12.$$

But we know $x = 4y$. So, by substitution, $2(4y) - 5y = 12$, or $3y = 12$, so $y = 4$ (Ben's present age). In 4 years, Ben will be $y + 4 = 8$ years old.

Fraction Problems

Sometimes, trial and error finds the answer for you. Simply substitute the answer choices into the problem to see what works. But you should also be able to set up simultaneous equations. You might ask, "What is the best method?" The answer is, "Whatever gives you the correct answer the fastest." Everyone is different.

Example 8:   A fraction when reduced is $\frac{3}{4}$. If 1 is subtracted from the numerator and 2 is added to the denominator, the ratio becomes $\frac{2}{3}$. What is the original fraction?

Solution: We have $\frac{x}{y} = \frac{3}{4}$, or $3y = 4x$, which gives us $-4x + 3y = 0$. We also have $\frac{x-1}{y+2} = \frac{2}{3}$, or, by cross-multiplication, $3(x-1) = 2(y+2)$. This gives us $3x - 2y = 7$. Our simultaneous equations are

$$-4x + 3y = 0$$
$$3x - 2y = 7$$

If we eliminate x by multiplying the top equation by 3 and the bottom equation by 4 and adding the resulting equations, we get $y = 28$, and by substitution, $x = 21$. So the original fraction is $\frac{21}{28}$.

Let's do a few exercises.

Exercise 6:   In a two-digit number, the tens digit is the square of the units digit. The difference between the number and the number reversed is 54. The original number is

A. 24   D. 71

B. 39   E. 93

C. 42

Exercise 7:   How many ounces of 40% alcohol must be mixed with 10 ounces of 70% alcohol to give a solution that is 45% alcohol?

A. 20   D. 50

B. 30   E. 60

C. 40

Exercise 8:   May's age is twice Fay's age. In 15 years, Fay will be $\frac{3}{5}$ as old as May. The sum of their original ages is

A. 66   D. 96

B. 75   E. 105

C. 90

Exercise 9:   If 4 less than x is the same as 7 more than the product of 4 and y, which answer choice is true?

A. $x + 4y - 11 = 0$ D. $x + 4y - 11 = 0$

B. $x + y + 15 = 0$ E. $4y - x + 11 = 0$

C. $x + y + 7 = 0$

Exercise 10:   The product of 4 and the sum of x and y is at least as large as the quotient of a divided by b. This can be written as

A. $4x + y - \dfrac{a}{b} \geq 0$ D. $4(x + y) + \dfrac{a}{b} > 0$

B. $4x + y - \dfrac{a}{b} > 0$ E. $\dfrac{a}{b} - 4x + 4y < 0$

C. $4(x + y) - \dfrac{a}{b} \geq 0$

Exercise 11:   The sum of the digits of a two-digit number is 10. The number reversed is 18 more than the original number. The original number is

A. 19   D. 46

B. 28   E. 55

C. 37

 Let's look at the answers.

Answer 6:   E: Trial and error is best for this problem. Of the answer choices, the only possible ones are 42 since $2^2 = 4$, and 93 since $3^2 = 9$. However, only for 93 is the second criterion true: $93 - 39 = 54$.

Answer 7: **D:** This is a mixture problem. The principle is that if we have 10 ounces of 70% alcohol, the amount of alcohol is 10(.70) = 7 ounces. Now, we construct a chart for this problem (we can eliminate the decimal points, since all terms have a two-place decimal).

	Ounces of each	× % alcohol =	Amount of Alcohol
A	x	40	40x
B	10	70	700
Mixture	y	45	45y

We let x = the amount of 40% alcohol, and y = the total ounces in the mixture. We get $x + 10 = y$ and $40(x) + 10(70) = 45(y)$. Substitute the first equation for y because we want x: $40x + 700 = 45(x + 10)$, or $45x + 450 = 40x + 700$. Thus, $5x = 250$, and $x = 50$.

Answer 8: **C:** Consruct a chart in which May's age is y and Fay's age is x.

	Age now	Age in 15 years
May	x	$y + 15$
Fay	y	$x + 15$

From the chart $y = 2x$ and $x + 15 = \left(\dfrac{3}{5}\right)(y + 15)$, or $5(x + 15) = 3(y + 15)$. Then $5x + 30 = 3y$. Substituting $y = 2x$, we get $5x + 30 = 3(2x)$; so $x = 30$ and $y = 60$. The sum of their ages is $x + y = 90$.

Answer 9: **E:** $x - 4 = 4y + 7$. When we rearrange the terms, we see that only E is correct.

Answer 10: **C:** $4(x + y) \geq \dfrac{a}{b}$. When we rearrange the terms, we see that only C is correct.

Answer 11: **D:** $t + u = 10$ and $10u + t = 10t + u + 18$, or $9u - 9t = 18$. Dividing both sides of this equation by 9, we get $u - t = 2$. We thus have $t + u = 10$ and $u - t = 2$. Adding these equations, we get $2u = 12$, so $u = 6$ and $t = 4$. Thus, the number is 46. We could have tried trial and error for this problem.

Let's take a break from algebra and word problems now, and look at some familiar lines and shapes.

"*Your journey began from a single point. You travel in a straight line; sometimes the slope may be steep and the distance seems far, but you are now at the midpoint. The endpoint is in sight.*"

This topic used to be part of a course called analytic geometry (algebraic geometry). We'll start at the beginning.

POINTS IN THE PLANE

We start with a **plane**—a two-dimensional space, like a piece of paper. On this plane, we draw two perpendicular lines, or **axes**. The x-axis is horizontal; the y-axis is vertical. Positive x is to the right; negative x is to the left. Positive y is up; negative y is down. Points in the plane are indicated by **ordered pairs** (x, y). The x number, called the **first coordinate** or **abscissa**, is always given first; the y number, called the **second coordinate** or **ordinate**, is always given second. Here are some points on the plane.

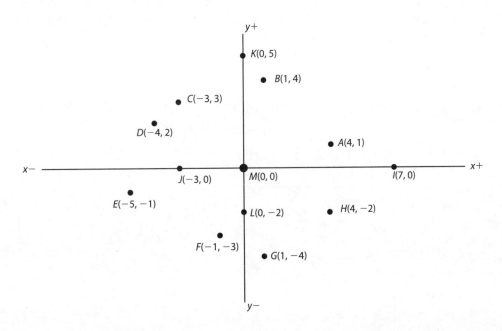

Note the following:

- For any point on the *x*-axis, the *y* coordinate always is 0.

- For any point on the *y*-axis, the *x* coordinate is 0.

- The point where the two axes meet, (0, 0), is called the **origin**.

- The axes divide the plane into four **quadrants**, usually written with roman numerals, starting in the upper right quadrant and going counterclockwise.

  - In quadrant I, $x > 0$ and $y > 0$.

  - In quadrant II, $x < 0$ and $y > 0$.

  - In quadrant III, $x < 0$ and $y < 0$.

  - In quadrant IV, $x > 0$ and $y < 0$.

$$
\begin{array}{c|c}
\text{II} & \text{I} \\
x < 0 & x > 0 \\
y > 0 & y > 0 \\
\hline
\text{III} & \text{IV} \\
x < 0 & x > 0 \\
y < 0 & y < 0 \\
\end{array}
$$

In the following figure, we have drawn the line $y = x$. For every point on this line, the first coordinate has the same value as the second coordinate, or $y = x$. If we shade the area above this line, $y > x$ in the shaded portion. Similarly, $x > y$ in the unshaded portion. Sometimes, questions on the GMAT ask about this.

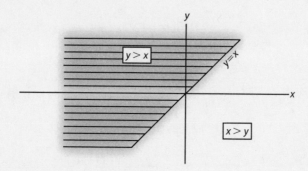

The following figure shows symmetry about the *x*-axis, *y*-axis, and the origin. Suppose (*a, b*) is in quadrant I. Then (−*a, b*) would be in quadrant II, (−*a,* −*b*) would be in quadrant III, and (*a,* −*b*) would be in quadrant IV, as pictured.

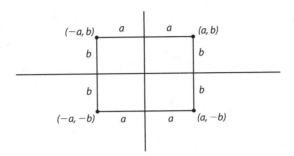

LINES

Distance and Midpoint

The formulas for distance and midpoint look a little complicated, but they are fairly easy to use. It just takes practice.

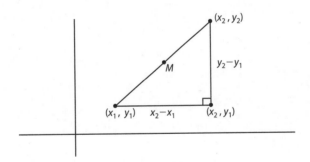

To find the **distance** between two points (x_1, y_1) and (x_2, y_2) on a plane, we must use the distance formula:

$$d = \sqrt{(x_2 - x_1)^2 + (y_2 - y_1)^2}$$

The distance formula is just the Pythagorean Theorem (discussed in the next chapter).

Distances are always positive. You may be six feet tall, but you cannot be minus six feet tall.

The **midpoint** of a line between two points (x_1, y_1) and (x_2, y_2) on a plane is given by

$$M = \left(\frac{x_1 + x_2}{2}, \frac{y_1 + y_2}{2} \right)$$

In one dimension, if the line is horizontal, these formulas simplify to $d = x_2 - x_1$ and $M = \dfrac{x_1 + x_2}{2}$.

Similarly, if the line is vertical, these formulas simplify to $d = y_2 - y_1$ and $M = \dfrac{y_1 + y_2}{2}$.

For example, for the horizontal line shown in the figure below, the distance between the points is $d = x_2 - x_1 = 7 - (-3) = 10$, and the midpoint is $M = \dfrac{x_1 + x_2}{2} = \dfrac{-3 + 7}{2} = 2$.

Similarly, for the vertical line shown in the figure below, the distance between the points is $d = y_2 - y_1 = -3 - (-7) = 4$, and the midpoint is $M = \dfrac{y_1 + y_2}{2} = \dfrac{(-7) + (-3)}{2} = -5$.

Slope

The **slope** of a line tells by how much the line is "tilted" compared to the x-axis. The formula for the slope of a line is

$$m = \frac{\text{rise}}{\text{run}} = \frac{\text{change in } y}{\text{change in } x} = \frac{y_2 - y_1}{x_2 - x_1},$$

where (x_1, y_1) and (x_2, y_2) are any two points on the line.

Note the following facts about the slope of a line, as shown in the figure below:

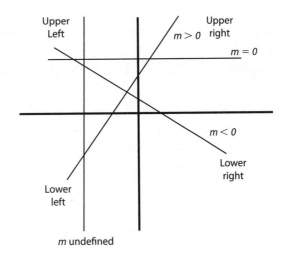

- The slope is positive if the line goes from the lower left to the upper right.
- The slope is negative if it goes from the upper left to the lower right.
- Horizontal lines have zero slope.
- Vertical lines have no slope or undefined slope or "infinite" slope.

Example 1: Find the distance, slope, and midpoint for the line segment joining these points:

a. (2, 3) and (6, 8) **c.** (7, 3) and (4, 3)

b. (4, −3) and (−2, 0) **d.** (2, 1) and (2, 5)

Solutions:

a. We let $(x_1, y_1) = (2, 3) =$ and $(x_2, y_2) = (6, 8)$, although the other way around is also okay.

Then

$$\text{Distance} = d = \sqrt{(x_2 - x_1)^2 + (y_2 - y_1)^2} = \sqrt{(6-2)^2 + (8-3)^2} = \sqrt{41}$$

$$\text{Slope} = m = \frac{y_2 - y_1}{x_2 - x_1} = \frac{8-3}{6-2} = \frac{5}{4}$$

$$\text{Midpoint} = M = \left(\frac{x_1 + x_2}{2}, \frac{y_1 + y_2}{2}\right) = \left(\frac{2+6}{2}, \frac{3+8}{2}\right) = (4, 5.5)$$

Notice that the slope is positive; the line segment goes from the lower left to the upper right.

b. We let $(x_1, y_1) = (4, -3) =$ and $(x_2, y_2) = (-2, 0)$.

Distance $= d = \sqrt{(-2 - 4)^2 + (0 - (-3))^2} = \sqrt{45}$

$$= \sqrt{3 \times 3 \times 5} = 3\sqrt{5}$$

Slope $= m = \dfrac{0 - (-3)}{-2 - 4} = -\dfrac{1}{2}$

Midpoint $= M = \left(\dfrac{4 + (-2)}{2}, \dfrac{-3 + 0}{2} \right) = (1, -1.5)$

Notice that the slope is negative; the line segment goes from the upper left to the lower right.

c. We let $(x_1, y_1) = (7, 3) =$ and $(x_2, y_2) = (4, 3)$.

It is a one-dimensional distance, so $d = |4 - 7| = 3$

Slope $= m = \dfrac{3 - 3}{4 - 7} = \dfrac{0}{-3} = 0$

Midpoint $= M = \left(\dfrac{7 + 4}{2}, \dfrac{3 + 3}{2} \right) = (5.5, 3)$

Notice that the horizontal line segment has slope $m = 0$.

d. We let $(x_1, y_1) = (2, 1) =$ and $(x_2, y_2) = (2, 5)$.

Again, this is a one-dimensional distance, so $d = |5 - 1| = 4$

Slope $= m = \dfrac{5 - 1}{2 - 2} = \dfrac{4}{0}$, undefined

Midpoint $= M = \left(\dfrac{2 + 2}{2}, \dfrac{5 + 1}{2} \right) = (2, 3)$

Notice that the slope of the vertical line segment is undefined.

Q **Now let's do some exercises.**

Exercise 1:   The coordinates of *P* are (*j, k*). If s < *k* < *j* < *r*, which of the points shown in the figure could have the coordinates (*r, s*)?

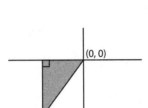

A. A

D. D

B. B

E. E

C. C

Use the figure below for Exercises 2 and 3.

Exercise 2:   Which of the following points is inside the triangle?

A. (−3, 6)

D. (−3, −4)

B. (−5, −5)

E. (−1, −3)

C. (−2, −5)

Exercise 3:   The area of the triangle is

A. 6

D. 24

B. 12

E. 48

C. 18

Exercise 4:   *M* is the midpoint of line segment *AB*. If the coordinates of *A* are (*m, −n*), then the coordinates of *B* are

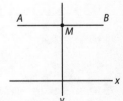

A. (*m, n*)

D. (*n, m*)

B. (−*m, n*)

E. (−*n, −m*)

C. (−*m, −n*)

Exercise 5:   In the given figure, $AB \parallel x$-axis and $PQ = AB$.
The coordinates of point A are

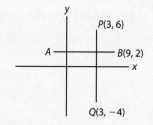

A. $(-1, 2)$ **D.** $(9, 0)$

B. $(1, 2)$ **E.** $(-9, 2)$

C. $(9, -8)$

 Let's look at the answers.

Answer 1:   C: For points C, D, and E, the x value is bigger than the x value of P; only point C has a y value less than the y value of P.

Answer 2:   D.

Answer 3:   B: We really haven't gotten to this, but I asked it because we have the picture. The area of the triangle is half the area of the rectangle.

$$A = \frac{1}{2}bh = \frac{1}{2} \times 4 \times 6 = 12$$

Answer 4:   C: Slightly tricky. Point B has the same y value as A, but its x value is the negative of the x value for A.

Answer 5:   A: The length of $PQ = 10$. For the length of AB to be 10, A must be $(-1, 2)$ since $9 - (-1) = 10$.

Standard Equation of a Line

Let's go over the facts we need.

- **Standard form** of the line: $Ax + By = C$; A, B both $\neq 0$.

- The **x-intercept**, the point at which the line hits the x-axis, occurs when $y = 0$.

- The **y-intercept**, the point at which the line hits the y-axis, occurs when $x = 0$.

- **Point-slope form** of a line: Given slope m and point (x_1, y_1), the point-slope form of a line is $m = \dfrac{y - y_1}{x - x_1}$.

- **Slope-intercept form** of a line: $y = mx + b$, where m is the slope and $(0, b)$ is the y-intercept.

- Lines of the form:

  y = constant, such as $y = 2$, are lines parallel to the x-axis; the equation of the x-axis is $y = 0$.

  x = constant, such as $x = -3$, are lines parallel to the y-axis; the equation of the y-axis is $x = 0$.

  $y = mx$ are lines that pass through the origin.

  Example 2:   For $Ax + By = C$, find the x and y intercepts.

  Solution: The y-intercept means $x = 0$; so $y = \dfrac{C}{B}$, and the y intercept is $\left(0, \dfrac{C}{B}\right)$.

  The x-intercept means $y = 0$; so $x = \dfrac{C}{A}$, and the x intercept is $\left(\dfrac{C}{A}, 0\right)$.

  Example 3:   For $3x - 4y = 7$, find the x and y intercepts.

  Solution: For the y-intercept, $x = 0$; so $y = \dfrac{7}{-4}$, and the y-intercept is $\left(0, -\dfrac{7}{4}\right)$. For the x-intercept, $y = 0$; $x = \dfrac{7}{3}$, and the x-intercept is $\left(\dfrac{7}{3}, 0\right)$.

  Example 4:   Given $m = \dfrac{3}{2}$ and point $(5, -7)$, write the equation of the line in standard form.

  Solution: $m = \dfrac{y - y_1}{x - x_1}$, so $\dfrac{3}{2} = \dfrac{y - (-7)}{x - 5}$. Cross-multiplying, we get $3(x - 5) = 2(y + 7)$, or $3x - 2y = 29$.

  Example 5:   Given points $(3, 6)$ and $(7, 11)$, write the line in slope-intercept form.

  Solution: $y = mx + b$. $m = \dfrac{11 - 6}{7 - 3} = \dfrac{5}{4}$, and we will use point $(3, 6)$, so $x = 3$ and $y = 6$. Therefore, $6 = \dfrac{5}{4}(3) + b$, and $b = \dfrac{9}{4}$. So the line is $y = \dfrac{5}{4}x + \dfrac{9}{4}$.

  Example 6:   Sketch lines $x = -3$, $y = 8$, and $y = \dfrac{2}{3}x$.

  Solution:

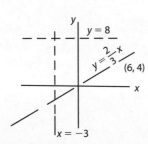

Q   **Let's do a few more exercises.**

Exercise 6:   A line with the same slope as the line $y = \frac{2}{3}x - 5$ is

        A. $2x = 18 - 3y$ **D.** $-2x - 3y = 14$

        B. $2x + 3y = 6$ **E.** $2y = 12 - 3x$

        C. $2x - 3y = 6$

Exercise 7:   Find the area of the triangle formed with the positive *x*-axis, positive *y*-axis, and the line though the point (3, 4) with slope −2. The area is

        A. 5   **D.** 50

        B. 15   **E.** 10

        C. 25

A   **Let's look at the answers.**

Answer 6:   **C:** We have to solve for *y* in each case, but we are interested in only the coefficient of *x*. The only answer choice that works is C.

Answer 7:   **C:** You must visualize the figure.

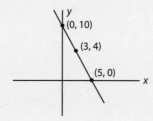

The area of the triangle is one-half the *x*-intercept times the *y*-intercept.

The equation of the line is $-2 = \frac{y - 4}{x - 3}$. If we let $x = 0$, the *y*-intercept

is 10. If we let $y = 0$, the *x*-intercept is 5. Area $= \frac{1}{2}ab = \frac{1}{2} \times 5 \times 10 = 25$.

Let's finally get to angles and triangles.

CHAPTER 10: *About Angles and Triangles*

"Understanding the area and its perimeter will enhance your chances for success. **"**

Before I wrote this chapter, I formulated in my head how the chapter would go. Too many of the questions on angles had to do with triangles. So I decided to write the chapters together. Let's start with some definitions.

TYPES OF ANGLES

There are several ways to classify angles, such as by angle measure, as shown here:

Acute angle: An angle of less than 90°.

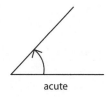

acute

Right angle: A 90° angle. As we will see, some other words that indicate a right angle or angles are perpendicular ($\perp$), altitude, and height.

right

Obtuse angle: An angle of more than 90° but less than 180°.

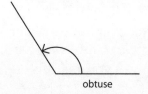

obtuse

Straight angle: An angle of 180°.

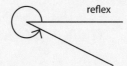

Reflex angle: An angle of more than 180° but less than 360°.

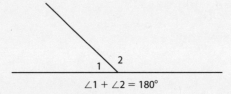

Angles are also named for their relation to other angles, such as:

Supplementary angles: Two angles that total 180°. $\angle 1 + \angle 2 = 180°$.

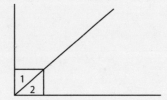

Complementary angles: Two angles that total 90°.

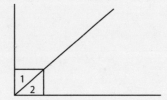

 A note of interest: Once around a circle is 360°. The reason that it is 360° is that the ancient Babylonians, about 7000 years ago, thought there were 360 days in a year. Three hundred sixty degrees is unique to the planet Earth.

You probably learned that angles are congruent and measures of angles are equal. I am using what I learned; it is simpler and makes understanding easier. So "angle 1 equals angle 2" (or $\angle 1 = \angle 2$) means the angles are both congruent and equal in degrees.

ANGLES FORMED BY PARALLEL LINES

Let's look at angles formed when a line crosses two parallel lines. In the figure below, $\ell_1 \| \ell_2$, and *t* is a **transversal**, a line that cuts two or more lines. It is not important that you know the names of these angles, although many of you will. It is important only to know that angles formed by a line crossing parallel lines that look equal are equal. The angles that are not equal add to 180°. In this figure, $\angle 1 = \angle 4 = \angle 5 = \angle 8$ and $\angle 2 = \angle 3 = \angle 6 = \angle 7$. Any angle from the first group added to any angle from the second group totals 180°.

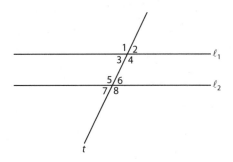

Vertical angles, which are the opposite angles formed when two lines cross, are equal. In the figure below, $\angle 1 = \angle 3$ and $\angle 2 = \angle 4$. Also, $\angle 1 + \angle 2 = \angle 2 + \angle 3 = \angle 3 + \angle 4 = \angle 4 + \angle 1 = 180°$.

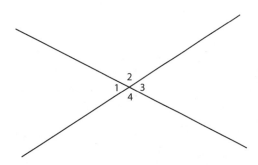

Q **Let's do some exercises.**

Exercise 1: $\angle b =$

    A. 45°   D. 105°

    B. 60°   E. 135°

    C. 90°

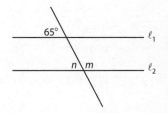

Exercise 2: $\ell_1 \parallel \ell_2.\ m - n =$

    A. 30°   D. 90°

    B. 50°   E. 180°

    C. 65°

Exercise 3: $y + z =$

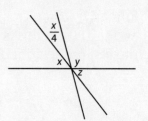

A. $180° - x$

D. $90° + \dfrac{5x}{4}$

B. $180° - \dfrac{x}{4}$

E. $90° - \dfrac{5x}{4}$

C. $45° - \dfrac{x}{4}$

Exercise 4: $180° - w =$

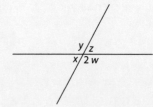

A. $x + w$

D. $y - z$

B. $x + y$

E. $z - w$

C. $y + z$

Exercise 5: $b =$

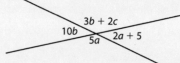

A. $5.5°$

D. $12.5°$

B. $7°$

E. Cannot be
    determined

C. $10°$

Exercise 6: y (in terms of x) $=$

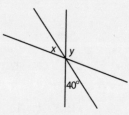

A. x

D. $140° + x$

B. $x + 40°$

E. $320° - x$

C. $140° - x$

Exercise 7: $\angle x =$

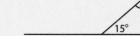

A. $70°$

D. $290°$

B. $110°$

E. $345°$

C. $210°$

Exercise 8:   The ratio of $a°$ to $(a + b)°$ is 3 to 8; $a =$

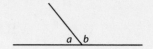

A. $60°$

D. $108°$

B. $67.5°$

E. $112.5°$

C. $72°$

Now let's look at the answers.

Answer 1:   **A:** $3a + a = 180°$; $a = 45°$ and $b = a = 45°$.

Answer 2:   **B:** $n = 65°$ and $n + m = 180°$; so $m = 115°$, and $m - n = 50°$.

Answer 3:   **B:** $\dfrac{x}{4} + y + z = 180°$, so $y + z = 180° - \dfrac{x}{4}$.

Answer 4:   **A:** Below the line, $x + 2w = x + w + w = 180°$, so $x + w = 180° - w$.

Answer 5:   **A:** This is a toughie. Don't look at vertical angles, look at the supplementary angles. On the bottom, we have $5a + 2a + 5° = 180°$, so $7a = 175°$, and $a = 25°$. Then, on the left, $10b + 5a = 180°$. Substituting $a = 25°$, we get $10b = 180° - 125° = 55°$, or $b = 5.5°$.

We could also have looked at the vertical angles, once we determined that $a = 25°$. Then $10b = 2a + 5° = 2(25°) + 5° = 55°$, so $b = 5.5°$.

Answer 6:   **C:** $x + y + 40° = 180°$; so $y = 140° - x$.

Answer 7:   **D:** Draw $\ell_3 \parallel \ell_1$ and ℓ_2.

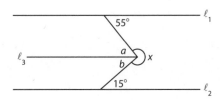

$\angle x = 360° - 70° = 290°$.

Answer 8:   **B:** One way to answer this exercise is to say $a = \left(\dfrac{3}{8}\right) \times 180$. You will notice that if you divide 8 into 180, you will have a fraction (or a decimal). So only answer choices B or E could be correct. Since B is < 90, B must be the correct answer. Using logic on the GMAT could save you a lot of time and give you more correct answers quickly.

So many angle questions on the GMAT involve triangles that we ought to look at triangles next.

TRIANGLES

Basics about Triangles

A **triangle** is a polygon with three sides. Angles are usually indicated with capital letters. The side opposite the angle is indicated with the same letter, only lowercase.

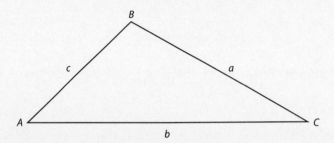

You should know the following general facts about triangles.

- The **sum of the angles** of a triangle is 180°.

- The **altitude**, or **height** (*h*), of △*ABC* shown below is the line segment drawn from a vertex perpendicular to the base, extended if necessary. The **base** of the triangle is *AC* = *b*.

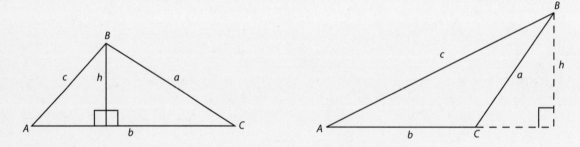

- The **perimeter of a triangle** is the sum of the three sides: $p = a + b + c$.

- The **area of a triangle** is $A = \dfrac{1}{2}bh$. The reason is that a triangle is half a rectangle. Since the area of a rectangle is base times height; a triangle is half a rectangle, as shown in the figure below.

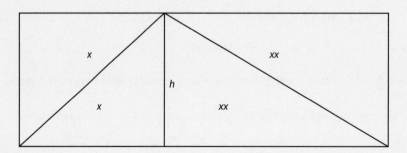

- An **angle bisector** is a line that bisects an angle in a triangle. In the figure below, *BD* bisects ∠*ABC* if ∠1 = ∠2.

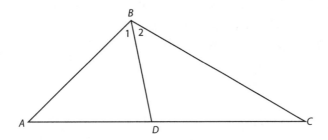

- A **median** is a line drawn from any angle of a triangle to the midpoint of the opposite side. In the figure below, *BD* is a median to side *AC* if *D* is the midpoint of *AC*.

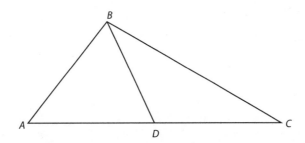

There are many kinds of triangles. One way to describe them is by their sides.

- A **scalene** triangle has three unequal sides and three unequal angles.

- An **isosceles** triangle has at least two equal sides. In the figure below, side *BC* (or *a*) is called the **base**; it may be equal to, greater than, or less than any other side. The **legs**, *AB* = *AC* (or *b* = *c*) are equal. Angle *A* is the **vertex angle**; it may equal the others, or be greater than or less than the others. The **base angles** are equal: ∠*B* = ∠*C*.

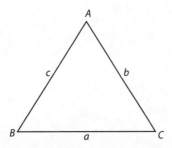

- An **equilateral** triangle is a triangle with all equal sides. All angles equal 60°, so this triangle is sometimes called an **equiangular** triangle. For an equilateral triangle of side *s*, the perimeter *p* = 3*s*, and the area $A = \dfrac{s^2\sqrt{3}}{4}$. This formula seems to be very popular lately, and you may see it on the GMAT.

Triangles can also be described by their angles.

- An **acute** triangle has three angles that are less than 90°.

- A **right** triangle has one right angle, as shown in the figure below. The **right angle** is usually denoted by the capital letter C. The **hypotenuse** AB is the side opposite the right angle. The **legs**, AC and BC, are not necessarily equal. ∠A and ∠C are always **acute** angles.

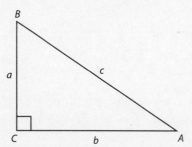

- An **obtuse** triangle has one angle between 90° and 180°.

An **exterior angle** of a triangle is formed by extending one side. In the figure below, ∠1 is an exterior angle. An exterior angle equals the sum of its two remote interior angles: ∠1 = ∠A + ∠B.

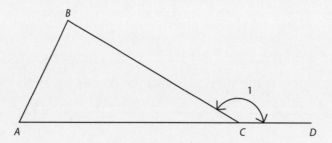

There are two other facts about triangles you should know:

1. The sum of any two sides of a triangle must be greater than the third side.

2. The largest side lies opposite the largest angle; and the largest angle lies opposite the largest side, as shown in the figure below.

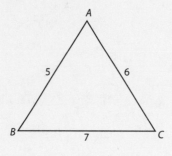

$$\angle C < \angle B < \angle A$$

Example 1: Give one set of angles for a triangle that satisfies the following descriptions:

Description	**Solution**
a. Scalene, acute	50°, 60°, 70°
b. Scalene, right	30°, 60°, 90°: We will deal with this one soon.
c. Scalene, obtuse	30°, 50°, 100°
d. Isosceles, acute	20°, 80°, 80°
e. Isosceles, right	Only one: 45°, 45°, 90°: We will deal with this one soon also.
f. Isosceles, obtuse	20°, 20°, 140°
g. Equilateral	Only one: three 60° angles

Let's first do some exercises with angles. Then we'll turn to area and perimeter exercises. We'll finish the chapter with our old friend Pythagoras and his famous theorem.

 Let's do some more exercises.

Exercise 9: Two sides of a triangle are 4 and 7. If only integer measures are allowed for the sides, the third side must be taken from which set?

 A. {5, 6, 7, 8, 9, 10, 11}   D. {3, 4, 5, 6, 7, 8, 9, 10, 11}

 B. {4, 5, 6, 7, 8, 9, 10}   E. {1, 2, 3, 4, 5, 6, 7, 8, 9, 10, 11}

 C. {3, 4, 5, 6, 7, 8, 9, 10}

Exercise 10: Arrange the sides in order, largest to smallest, for the figure shown below.

 A. $a > b > c$ D. $b > c > a$

 B. $a > c > b$ E. $c > a > b$

 C. $b > a > c$

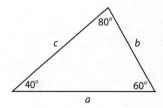

Exercise 11: $x = 2y; z =$

    A. 30°   **D.** 60°

    B. 40°   **E.** 90°

    C. 50°

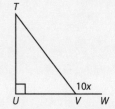

Exercise 12: WX bisects $\angle ZXY; \angle Z =$

    A. 20°   **D.** 60°

    B. 40°   **E.** 70°

    C. 50°

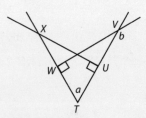

Exercise 13: $\angle TVW = 10x; x$ could be

    A. 3°   **D.** 16°

    B. 6°   **E.** 20°

    C. 9°

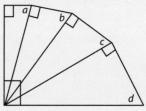

Exercise 14: Write b in terms of a:

    A. $a + 90°$ **D.** $180° - a$

    B. $2a$ **E.** $180° - 2a$

    C. $2a + 90°$

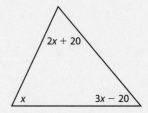

Exercise 15: $a + b + c + d =$

    A. 90°   **D.** 360°

    B. 180°   **E.** 450°

    C. 270°

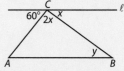

Exercise 16: The largest angle is

    A. 30°   **D.** 80°

    B. 50°   **E.** 90°

    C. 70°

Exercise 17: $\ell_1 \parallel AB; y =$

    A. 40°   **D.** 80°

    B. 60°   **E.** Can't be determined

    C. 70°

Exercise 18: $\ell_1 \parallel AB$; $y =$

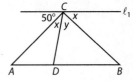

 A. 40°

 B. 60°

 C. 70°

 D. 80°

 E. Can't be determined

Use △ABC for Exercises 19 and 20.

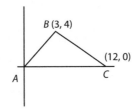

Exercise 19: The area of △ABC is

 A. 18

 B. 24

 C. 36

 D. 48

 E. 60

Exercise 20: The perimeter of △ABC is

 A. $17 + \sqrt{97}$

 B. 27

 C. 32

 D. $\sqrt{266}$

 E. $10\sqrt{10}$

For Exercises 21 and 22, use this figure of a square with an equilateral triangle on top of it, $AE = 20$.

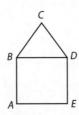

Exercise 21: The perimeter of ABCDE is

 A. 50

 B. 100

 C. 120

 D. 160

 E. 200

Exercise 22:   The area of *ABCDE* is

  A. 600   **D.** 800

  B. $100(4 + \sqrt{2})$ **E.** 1,000

  C. $100(4 + \sqrt{3})$

For Exercises 23 and 24, use △*ABC* with midpoints *X*, *Y*, and *Z*.

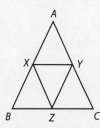

Exercise 23:   If the perimeter of △*ABC* is 1, the perimeter of △*XYZ* is

  A. $\dfrac{1}{16}$ **D.** $\dfrac{1}{2}$

  B. $\dfrac{1}{8}$ **E.** 1

  C. $\dfrac{1}{4}$

Exercise 24:   If the area of △*ABC* is 1, the area of △*XYZ* is

  A. $\dfrac{1}{16}$ **D.** $\dfrac{1}{2}$

  B. $\dfrac{1}{8}$ **E.** 1

  C. $\dfrac{1}{4}$

Exercise 25:   In the figure shown, $BC = \dfrac{1}{3}BD$. If the area of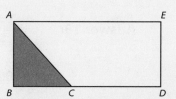
△*ABC* = 10, the area of rectangle *ABDE* is

  A. 30   **D.** 120

  B. 40   **E.** Can't be
                                              determined

  C. 60

Let's look at the answers.

Answer 9:   **B:** The third side s must be greater than the difference and less than the sum of the other two sides, or $> 7 - 4$ and $< 7 + 4$. Thus the third side must be between 3 and 11.

Answer 10:   **B:** Judge the relative lengths of the sides by the size of the angles opposite them. Then $a > c > b$.

Watch out for the words "Not drawn to scale." If it is a simple figure, "not drawn to scale" usually means it is not drawn to scale, and you cannot assume relative sizes without being given actual measurements. However, if it is a semi-complicated or complicated figure, the figure probably *is* drawn to scale.

Answer 11:   **A:** $y = 30°$; $x = 60°$; and $z = 30°$

Answer 12:   **C:** $\angle ZXY = 40°$, so $\angle Z$ must be $50°$.

Answer 13:   **D:** $\angle TVW$ must be between $90°$ and $180°$, so $9° < x < 18°$.

Answer 14:   **A:** This is really tricky. UX is drawn to confuse you. In $\triangle TVW$, b is the exterior angle, so $b = a + 90°$.

Answer 15:   **C:** The sum of 4 triangles is $4 \times 180° = 720°$. The sum of 5 right angles (don't forget the one in the lower left of the figure, which is the sum of four acute angles of the triangles) is $450°$; so $a + b + c + d = 720° - 450° = 270°$.

Answer 16:   **D:** $x + 2x + 20 + 3x - 20 = 180$, or $6x = 180$, so $x = 30$. $2x + 20 = 80$ and $3x - 20 = 70$. The largest angle is $80°$.

Answer 17:   **A:** $2x + x + 60 = 180$; $x = 40°$. But $y = x = 40°$ (because $\ell_1 \parallel AB$).

Answer 18:   **E:** y cannot be determined. The GMAT occasionally asks a question for which there is no answer. However, I've never seen two in a row and I've seen thousands of similar questions.

Answer 19:   **B:** $A = \frac{1}{2}bh = \frac{1}{2} \times 12 \times 4 = 24$.

Answer 20:   **A:** Use the distance formula to find sides AB and BC. $p = AC + AB + BC = 12 + \sqrt{4^2 + 3^2} + \sqrt{(3 - 12)^2 + (4 - 0)^2} = 12 + \sqrt{25} + \sqrt{97} = 17 + \sqrt{97}$.

Answer 21:   **B:** Do not include *BD*; $p = 5 \times 20 = 100$.

Answer 22:   **C:** Area $= s^2 + \dfrac{s^2\sqrt{3}}{4} = 20^2 + \dfrac{20^2\sqrt{3}}{4} = 400 + 100\sqrt{3}$.

Answer 23:   **D:** If the perimeter of $\triangle ABC$ is 1, and all the sides of $\triangle XYZ$ are half of those of $\triangle ABC$, so is the perimeter.

Answer 24:   **C:** If the sides of $\triangle XYZ$ are half of those of $\triangle ABC$, the area of $\triangle XYZ$ is $\left(\dfrac{1}{2}\right)^2 A = \dfrac{1}{4}A = \dfrac{1}{4}$.

Answer 25:   **C:** If we draw lines parallel to *DE* to divide the original rectangle into three congruent rectangles, and then divide each rectangle into two triangles, we see that each triangle is one-sixth of the rectangle. So the area of the rectangle is $6(10) = 60$.

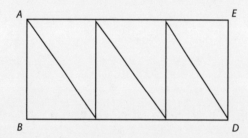

We'll have more of these type of exercises as part of Chapter 12, Circles.

Let's go on to good old Pythagoras.

PYTHAGOREAN THEOREM

This is perhaps the most famous math theorem of all. Most theorems have one proof. A small fraction of these have two. This theorem, however, has more than a hundred, including three by past presidents of the United States. We've had some smart presidents who actually knew some math.

The Pythagorean Theorem simply states:

In a right triangle, the hypotenuse squared is equal to the sum of the squares of the legs.

In symbols,

$$c^2 = a^2 + b^2.$$

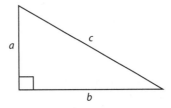

As a teacher, I must show you one proof.

Proof:

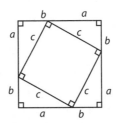

In this figure, the larger square equals the smaller square plus the four congruent triangles. In symbols, $(a + b)^2 = c^2 + 4\left(\frac{1}{2}ab\right)$.

Multiplying this equation out, we get $a^2 + 2ab + b^2 = c^2 + 2ab$. Then canceling $2ab$ from both sides, we get $c^2 = a^2 + b^2$. The proof is complete.

There are two basic problems you need to know how to do: finding the hypotenuse and finding one of the legs of the right triangle.

Example 2:　Solve for *x*:

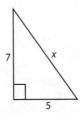

Solution:　$x^2 = 7^2 + 5^2; x = \sqrt{74}$.

Example 3:   Solve for *x*:

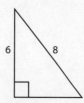

Solution: $8^2 = 6^2 + x^2$, or $x^2 = 64 - 36 = 28$. So $x = \sqrt{28} = \sqrt{2 \times 2 \times 7} = 2\sqrt{7}$.

Notice that the hypotenuse squared is always by itself, whether it is a number or a letter.

Pythagorean Triples

Because no calculator is allowed on the exam, it is a good idea to memorize some Pythagorean triples. These are the measures of sides of a triangle that are *always* right triangles. The hypotenuse is always listed third in the group.

The 3-4-5 group: 3-4-5, 6-8-10, 9-12-15, 12-16-20, 15-20-25

The 5-12-13 group: 5-12-13, 10-24-26

The rest: 8-15-17, 7-24-25, 20-21-29, 9-40-41, 11-60-61

Special Right Triangles

You ought to know two other special right triangles, the isosceles right triangle (with angles 45°-45°-90°) and the 30°-60°-90° right triangle. The facts about these triangles can all be found by using the Pythagorean Theorem.

1.   The 45°-45°-90° isosceles right triangle:

- The legs are equal.
- To find a leg given the hypotenuse, divide the hypotenuse by $\sqrt{2}$ (or multiply by $\frac{\sqrt{2}}{2}$).
- To find the hypotenuse given a leg, multiply the leg by $\sqrt{2}$.

Example 4:   Find *x* and *y* for this isosceles right triangle.

Solution: $x = 5$ (the legs are equal); $y = 5\sqrt{2}$.

Example 5: Find *x* and *y* for this isosceles right triangle.

Solution: $x = y = \dfrac{18}{\sqrt{2}} = 18 \times \dfrac{\sqrt{2}}{2} = 9\sqrt{2}$.

2. The 30°-60°-90° right triangle.

- If the shorter leg (opposite the 30° angle) is not given, get it first. It is always half the hypotenuse.
- To find the short leg given the hypotenuse: divide the hypotenuse by 2.
- To find the hypotenuse given the short leg: multiply the short leg by 2.
- To find the short leg given the long leg: divide the long leg by $\sqrt{3}$ (or multiply by $\dfrac{\sqrt{3}}{3}$).
- To find the long leg given the short leg: multiply the short leg by $\sqrt{3}$.

Example 6: Find *x* and *y* for this right triangle.

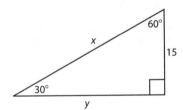

Solution: The short leg is given (15); $x = 2(15) = 30$; $y = 15\sqrt{3}$.

Example 7: Find *x* and *y* for this right triangle.

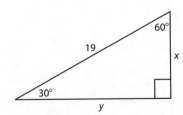

Solution: $x = \dfrac{19}{2} = 9.5.\ y = 9.5\sqrt{3}$.

Example 8:   Find x and y for this right triangle.

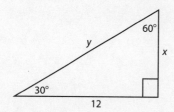

Solution: $x = \dfrac{12}{\sqrt{3}} = 12\dfrac{\sqrt{3}}{3} = 4\sqrt{3}$; $y = 2(4\sqrt{3}) = 8\sqrt{3}$.

Q   **Let's do a few exercises.**

Exercise 26:   Two sides of a right triangle are 3 and $\sqrt{5}$.

    I: The third side is 2.

    II: The third side is 4.

    III: The third side is $\sqrt{14}$.

    Which of the following choices is correct?

    A. Only II is true   **D.** Only I and III are true

    B. Only III is true   **E.** I, II, and III are true

    C. Only I and II are true

Exercise 27:   The area of square $ABCD =$

    A. 50   **D.** 576

    B. 100   **E.** 625

    C. 225

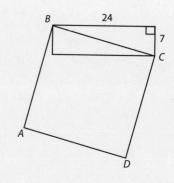

Exercise 28: $x =$

    A. 16   **D.** 22

    B. 18   **E.** 24

    C. 20

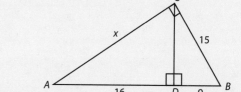

Exercise 29: $c^2 - b^2 =$

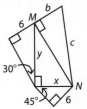

A. 72

D. 252

B. 144

E. 288

C. 216

Exercise 30: $x =$

A. 1

D. 4

B. 2

E. 4.5

C. 3

Exercise 31: A 25-foot ladder is leaning on the floor. Its base is 15 feet from the wall. If the ladder is pushed until it is only 7 feet from the wall, how much farther up the wall is the ladder pushed?

A. 4 feet

D. 20 feet

B. 8 feet

E. 24 feet

C. 12 feet

 Let's look at the answers.

Answer 26: D: Try the Pythagorean Theorem with various combinations of 3, $\sqrt{5}$, and x (the third side). The only ones that work are Statement I: $2^2 + \left(\sqrt{5}\right)^2 = 3^2$; and Statement III: $3^2 + \left(\sqrt{5}\right)^2 = \left(\sqrt{14}\right)^2$.

Answer 27: E: We recognize the right triangle as a 7-24-25 triple, so side $BC = 25$. The area of the square is $(25)^2 = 625$.

Answer 28: C: This is a 15-20-25 triple, so $x = 20$.

Answer 29: C: We see that MN is the side of two triangles. By the Pythagorean Theorem, we get $c^2 - b^2 = x^2 + y^2 = \left(6\sqrt{2}\right)^2 + 12^2 = 72 + 144 = 216$.

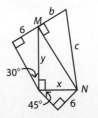

Answer 30:   **D:** This triangle is a 12-16-20 triple, so $3x + 2x = 5x = 20$, and $x = 4$.

Answer 31:   **A:** The first figure is shows a 15-20-25 right triangle with the ladder 20 feet up the wall. The second figure is a 7-24-25 triple with the ladder 24 feet up the wall. The ladder is pushed another $24 - 20 = 4$ feet up the wall.

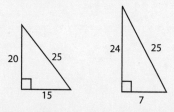

That's all for angles and triangles for now. We will see more when circles are discussed in Chapter 12. For now, though, let's look at rectangles and other polygons.

CHAPTER 11: *Quadrilaterals and Other Polygons*

"*Mastering all shapes and sizes will enhance your journey. We now deal with the rest of the polygons (closed figures with line-segment sides.*"

We now deal with the rest of the polygons (closed figures with line-segment sides).

QUADRILATERALS

Parallelograms

A **parallelogram** is a **quadrilateral** (four-sided polygon) with parallel opposite sides.

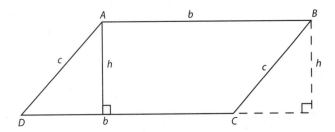

You should know the following properties about parallelograms:

- The opposite angles are equal. $\angle DAB = \angle BCD$ and $\angle ADC = \angle ABC$.
- The consecutive angles are supplementary. $\angle DAB + \angle ABC = \angle ABC + \angle BCD = \angle BCD + \angle CDA = \angle CDA + \angle DAB = 180°$.
- The opposite sides are equal. $AB = CD$ and $AD = BC$.
- The diagonals bisect each other. $AE = EC$ and $DE = EB$.

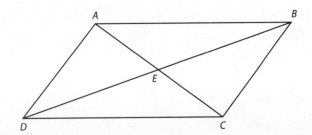

- Area $= A = bh$. This is a **postulate** (law taken to be true without proof) from which we get the area of all other figures with sides that are line segments.

- Perimeter $= p = 2b + 2c$

Example 1:   For parallelogram *RSTU*, find the following if $RU = 10$:

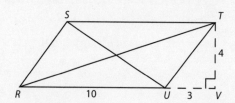

a. The area   c. Diagonal *RT*

b. The perimeter   d. Diagonal *SU*

Solutions: a. $A = bh = (10)(4) = 40$ square units. The whole test should be this easy!

b. We have to find the length of $RS = TU$. $TU = 5$ because it is the hypotenuse of a 3-4-5 right triangle. So the perimeter is
$p = 2(10) + 2(5) = 30$ units.

c. $RT = \sqrt{(RV)^2 + (TV)^2} = \sqrt{13^2 + 4^2} = \sqrt{185}$

d.

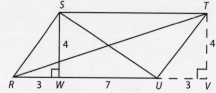

To find diagonal *SU*, draw the other altitude *SW* as pictured.

$SU = \sqrt{WU^2 + SW^2} = \sqrt{7^2 + 4^2} = \sqrt{65}$

Example 2: For parallelogram *WXYZ* with altitudes *XM* and *YN*, find the following in terms of *a*, *b*, and *c*:

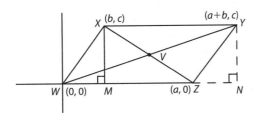

a. The coordinates of point *M*

b. The coordinates of point *N*

c. The coordinates of point *V*

d. The perimeter

e. The area

Solutions: a. *M* has the same *x*-coordinate as point *X* and the same *y*-coordinate as point *W*, so the coordinates of *M* are $(b, 0)$.

b. *N* has the same *x*-coordinate as point *Y* and the same *y*-coordinate as point *W*, so the coordinates of *N* are $(a + b, 0)$.

c. *V* is halfway between *W* and *Y*, so use the formula for the midpoint between $Y(a + b, c)$ and $W(0,0)$: Midpoint $V = \left(\dfrac{(a + b) - 0}{2}, \dfrac{c - 0}{2} \right) = \left(\dfrac{a + b}{2}, \dfrac{c}{2} \right)$.

d. $WZ = XY$ is length *a*. By the distance formula, $WX = ZY = \sqrt{(b - 0)^2 + (c - 0)^2} = \sqrt{b^2 + c^2}$. Therefore, the perimeter is $p = 2a + 2\sqrt{b^2 + c^2}$.

e. Area $= A =$ base $\times$ height $= ac$.

Example 3: For parallelogram *EFGH*, find the smaller angle.

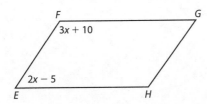

Solution: Consecutive angles of a parallelogram are supplementary. Therefore, $(3x+10)° + (2x - 5)° = 180°; x = 35°$; so the smaller angle is $2(35°) - 5° = 65°$. Be careful to give the answer the GMAT wants. Two other choices would be 35° and 115°, for those who do not read carefully!!!

Rhombus

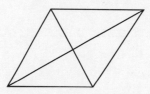

A **rhombus** is an equilateral parallelogram.

Thus, a rhombus has all of the properties of a parallelogram plus the following:

- All sides are equal.
- The diagonals are perpendicular bisectors of each other.
- Perimeter $= p = 4s$.
- Area $= A = bh = \frac{1}{2} \times d_1 \times d_2$, the area equals half the product of its diagonals.

Example 4: For the given rhombus with side $s = 13$ and larger diagonal $BD = 24$, find the other diagonal and the area.

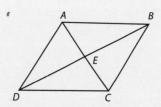

Solution: $AB = 13$ and $BD = 24$. Since the diagonals bisect each other, $BE = 12$.

The diagonals are perpendicular to each other, so $\triangle ABE$ is a 5-12-13 right

triangle, and $AE = 5$. Therefore, the other diagonal $AC = 10$. The area is

$A = \frac{1}{2} \times d_1 \times d_2 = \frac{1}{2}(24)(10) = 120$ square units.

Example 5: Find the area of a rhombus with side 10 and smaller interior angle of 60°.

Solution: If we draw the diagonal through the two larger angles, we will have two congruent equilateral triangles. The area of this rhombus is twice the area of each triangle, or $2 \times \dfrac{s^2\sqrt{3}}{4}$. Since $s = 10$, the area is

$$A = 2 \times \frac{10^2\sqrt{3}}{4} = 50\sqrt{3} \text{ square units.}$$

Now let's go on to more familiar territory.

Rectangle

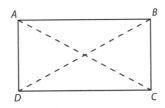

A **rectangle** is a parallelogram with right angles.

Therefore, it has all of the properties of a parallelogram plus the following:

- All angles are 90°.
- Diagonals are equal (but *not* perpendicular).
- Perimeter $= p = 2b + 2h$
- Area $= A = bh$

The easier the shape, the more likely the GMAT will have a problem or problems about it.

Example 6: One base of a rectangle is 8; and one diagonal is 9. Find all the sides and the other diagonal. Find the perimeter and area.

Solution: The top base and bottom base are both 8. Both diagonals are 9. The other two sides are each $\sqrt{9^2 - 8^2} = \sqrt{17}$. So the perimeter is $p = 16 + 2\sqrt{17}$ units; and the area is $A = 8\sqrt{17}$ square units.

Example 7: $AB = 10$, $BC = 8$, $EF = 6$, and $FG = 3$. Find the area of the shaded region of the figure.

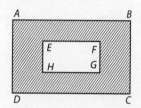

Solution: The area of the shaded region is the area of the outside rectangle minus the area of the inside one. $A = (10)(8) - (6)(3) = 62$ square units.

Example 8:   In polygon $ABCDEF$, $BC = 30$, $AF = 18$, $AB = 20$, and $CD = 11$. Find the perimeter and area of the polygon.

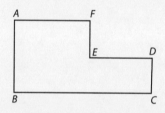

Solution: Draw a line through DE, hitting AB at point G. Then $AF = GE$ and $BC = GD$. Since $DG = 30$ and $GE = 18$, $DE = 12$. $AB = CD + EF$. $AB = 20$ and $CD = 11$, so $EF = 9$. This gives the lengths of all the sides. The perimeter thus is $p = AB + BC + CD + DE + EF + AF = 30 + 20 + 11 + 12 + 9 + 18 = 100$ units.

The area of rectangle $BCDG$ is $BC \times CD = (30)(11) = 330$. The area of rectangle $AFEG$ is $AF \times FE = (18)(9) = 162$. Therefore, the total area is $330 + 162 = 492$ square units. There are other ways to find this area, as you might be able to see.

Example 9: The areas of the pictured rectangle and triangle are the same.

If $\dfrac{LW}{4} = 20$, what is bh?

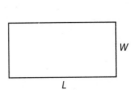 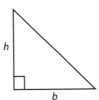

Solution: $\dfrac{LW}{4} = 20$; so $LW = 80$, which is the area of the rectangle. Since that also is the area of the triangle, $\dfrac{1}{2}bh = 80$. So $bh = 160$.

Square

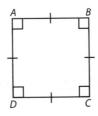

A **square** is a rectangle with equal sides, or it can be thought of as a rhombus with four equal 90° angles.

Therefore, it has all of the properties of a rectangle and a rhombus:

- All sides are equal.
- All angles are 90°.
- Both diagonals bisect each other, are perpendicular to each other, and are equal.
- Each diagonal $d = d_1 = d_2 = s\sqrt{2}$, where $s = $ a side.
- Perimeter $= p = 4s$
- Area $= A = \dfrac{d^2}{2} = s^2$.

Example 10: What is the area of this square?

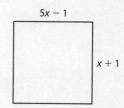

Solution: Since it is a square, $5x - 1 = x + 1$, so; $x = \dfrac{1}{2}$ and $x + 1 = 1\dfrac{1}{2}$. Then $A = \left(1\dfrac{1}{2}\right)^2 = 2\dfrac{1}{4}$.

Example 11: The area of square C is 36; the area of square B is 25. What is the area of square A?

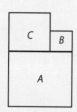

Solution: The side of square C must be 6, and the side of square B must be 5. Therefore, the side of square A is 11, and the area of square A is $11^2 = 121$.

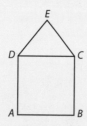

The figure is a square surmounted by an equilateral triangle. (I've always wanted to write that word.) $AB = 10$. Use this figure for Examples 12 and 13.

Example 12: What is the perimeter of the figure?

Solution: The perimeter is $5(10) = 50$. Note that CD is *not* part of the perimeter.

Example 13: What is the area of the figure?

Solution: $A = s^2 + \dfrac{s^2\sqrt{3}}{4} = 10^2 + \dfrac{10^2\sqrt{3}}{4} = 100 + 25\sqrt{3} = 25(4 + \sqrt{3})$.

Trapezoid

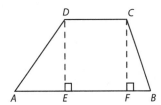

A **trapezoid** is a quadrilateral with exactly one pair of parallel sides.

Because a trapezoid is *not* a type of parallelogram, it has its own unique set of properties, as follows:

- The parallel sides, *AB* and *CD*, are called **bases**.
- The heights, *DE* and *CF*, are equal.
- The legs, *AD* and *BC*, may or may not be equal.
- The diagonals, *AC* and *BD*, may or may not be equal.
- Perimeter $= p = AB + BC + CD + AD$
- Area $= A = \dfrac{1}{2}h(b_1 + b_2)$, where b_1 and b_2 are the bases.

> **Note** *If we draw one of the diagonals, we see that a trapezoid is the sum of two triangles. Factoring out $\dfrac{1}{2}h$, we get the formula for the area of the trapezoid.*

If the legs are equal, the trapezoid is called an **isosceles trapezoid**, shown below.

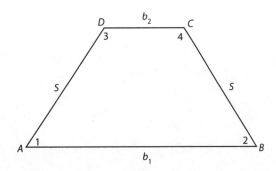

An isosceles trapezoid has the following additional properties:

- Perimeter $= p = b_1 + b_2 + 2s$
- The diagonals are equal, $AC = BD$.
- The base angles are equal, $\angle 1 = \angle 2$ and $\angle 3 = \angle 4$

Example 14: Find the area and the perimeter of Figure *ABCD*.

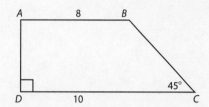

Solution: Draw the height from point *B*, *BG*, as shown.

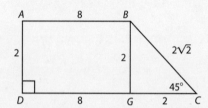

$DG = 8$; $CG = 2$; since $\triangle BGC$ is an isosceles right triangle, the height

BG (and AD) $= 2$. BC, the hypotenuse of the isosceles right triangle, is

therefore $2\sqrt{2}$. Therefore, the perimeter is $p = 10 + 2 + 8 + 2\sqrt{2} =$

$20 + 2\sqrt{2}$, and the area is $A = \dfrac{1}{2}h(b_1 + b_2) = \dfrac{1}{2}(2)(8 + 10) = 18$.

Example 15: Given trapezoid *ORST*, with *RS* ∥ *OT*, find the coordinates of point *S*. Find the perimeter and the area of trapezoid *ORST*.

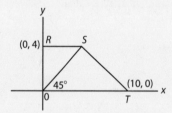

Solution: Since $\triangle ORS$ is a 45°-45°-90° triangle, $OR = RS = 4$, so *S* is the point

(4, 4). The length of $OT = 10$. By the distance formula, the length of

$ST = \sqrt{(10 - 4)^2 + (0 - 4)^2} = \sqrt{52} = 2\sqrt{13}$. Therefore, the perimeter is

$p = 10 + 4 + 4 + 2\sqrt{13}$, or, to be fancy, $2(9 + \sqrt{13})$. Area $= \dfrac{1}{2}h(b_1 + b_2)$

$= \dfrac{1}{2}(4)(10 + 4) = 28$. In a multiple-choice question, the GMAT would

ask about either the area or perimeter, but not both. But sometimes the

GMAT is fancy, like here.

Example 16: Find the area of isosceles trapezoid *EFGH*.

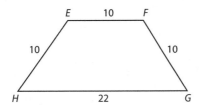

Solution: Draw in the two heights for the trapezoid.

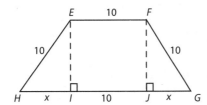

The two bases of the triangles formed are equal since it is an isosceles

trapezoid. From the figure, $2x + 10 = 22$, so $x = 6$. Each of the triangles

is a 6-8-10 Pythagorean triple, so the height of the trapezoid is 8.

Therefore, $A = \dfrac{1}{2}(8)(10 + 22) = 128$.

POLYGONS

Let's talk about polygons in general. Most of the time we deal with **regular** polygons. A regular polygon has all sides equal and all angles equal. A square and an equilateral triangle are examples of regular polygons we have already discussed.

Any *n*-sided polygon has the following properties.

- The sum of all the interior angles is $(n - 2)180°$, where *n* is the number of sides (or angles) in the polygon.

- The sum of all exterior angles always equals 360°.

- An interior angle plus its exterior angle always add to 180°.

- The number of diagonals is $\dfrac{n(n - 3)}{2}$.

If the polygon is regular, it has the following additional properties:

- Each exterior angle $= \dfrac{360°}{n}$.

- Each interior angle $= \dfrac{(n - 2)180°}{n}$.

Polygons are named for the number of sides they have.

A **pentagon** is a 5-sided polygon.

A **hexagon** is a 6-sided polygon.

A **heptagon** is a 7-sided polygon.

An **octagon** is an 8-sided polygon.

A **nonagon** is a 9-sided polygon.

A **decagon** is a 10-sided polygon.

A **dodecagon** is a 12-sided polygon.

An **n-gon** is an n-sided polygon.

Example 17: An octagon has a perimeter of 27. If 5 is added to each side, what is the perimeter of the new octagon?

Solution: It doesn't matter how long each side is! If 5 is added to each of 8 sides, 40 is added to the perimeter. The new perimeter is $27 + 40 = 67$.

Example 18: The side of a regular hexagon is 4. Find its area.

Solution: A regular hexagon is made up of six equilateral triangles. The side of each triangle is 4, and the area is $A = 6\dfrac{s^2\sqrt{3}}{4} = 6\dfrac{4^2\sqrt{3}}{4} = 24\sqrt{3}$.

Example 19: The sum of the interior angles of a polygon is 720°. Find the number of sides, the number of degrees in one exterior angle, and the number of degrees in one interior angle.

Solution: $(n - 2)(180) = 720$. Divide each side by 180 to simplify: $n - 2 = 4$, so $n = 6$ sides.

One exterior angle $= \dfrac{360°}{6} = 60°$. An interior angle is supplemental to its external angle, by definition, so $180° - 60° = 120°$ for each interior angle.

We'll get more GMAT-looking questions at the end of the chapter on circles, which is next.

"*Although sometimes it seems you are going in circles, you are really heading toward your goal.*"

Circles are a favorite topic of the GMAT. Circles allow for many short questions that can be combined with the other geometric chapters. Let's get started.

PARTS OF A CIRCLE

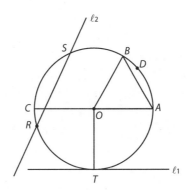

We all know what a circle looks like, but maybe we're not familiar with its "parts."

O is the **center** of the circle. *A* circle is often named by its center, so this is circle *O*.

OA, OT, OC, and *OB* are **radii** (singular: radius); a radius is a line segment from the center to the **circumference**, or edge, of the circle. *Note:* All radii (r) of a circle are equal. This is a postulate or axiom, a law taken to be true without proof. It is probably a good idea to tell you there are no proofs on the GMAT, as there probably were when you took geometry.

AC is the **diameter**, *d*, the distance from one side of the circle through the center to the other side; $d = 2r$ and $r = \dfrac{d}{2}$.

ℓ_1 is a **tangent**, a line that touches a circle in one and only one point

T is a **point of tangency**, the point where a tangent touches the circle. The radius to the point of tangency (*OT*) is always perpendicular to the tangent, so $OT \perp \ell_1$.

ℓ_2 is a **secant**, a line that passes through a circle in two places.

RS is a **chord**, a line segment that has each end on the circumference of the circle. The diameter is the longest chord in a circle.

OADBO is a **sector** (a pie-shaped part of a circle). There are a number of sectors in this figure; others include *BOCSB* and *OATRCSBO*. We will see these again soon.

An **arc** is any distance along the circumference of a circle.

> Arc *ADB* is a **minor arc** because it is less than half a circle. Arc *BDATRC* is a **major arc** because it is more than half a circle.

> Arc *ATRC* is a **semicircle** because it is exactly half a circle. Arc *ADBSC* is also a semicircle.

Whew! Enough! However, we do need some more facts about circles.

The following are mostly theorems, or proven laws. Again, there are no proofs on the GMAT, but you need to be aware of these facts.

AREA AND CIRCUMFERENCE

Area of a circle:

$$A = \pi r^2$$

Circumference (perimeter of a circle):

$$C = 2\pi r \text{ or } \pi d.$$

SECTORS

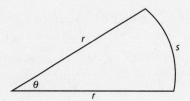

Area of a sector:

$$A = \frac{\theta}{360°}\pi r^2,$$

where θ (theta) is the angle of the sector in degrees.

Arc length of a sector:

$$s = \frac{\theta}{360°} \times 2\pi r$$

Perimeter of a sector:

$$p = s + 2r,$$

where *s* is the arc length

Example 1:   Find the area and perimeter of a 60° sector of a circle of diameter 12.

Solution:

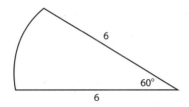

If the diameter is 12, the radius is 6. The sector is pictured here. Its area

is $A = \frac{60°}{360°} \times \pi 6^2 = 6\pi$ square units. Although you should know that

pi (π) is about 3.14, I've never seen a problem for which you had to

multiply 6 times 3.14. The answer is left in terms of π. The perimeter of

the sector is $s = 2(6) + \frac{60°}{360°} \times 2\pi(6) = 12 + 2\pi$ units.

Q   **Let's do some exercises.**

For Exercises 1 through 5, refer to the following circle, with center *O*.

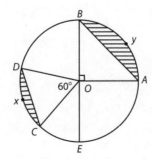

Exercise 1:   If the area of $\triangle AOB$ is 25, the area of circle O is

                  A. 12.5π **D.** 100π

                  B. 25π **E.** 200π

                  C. 50π

Exercise 2:   If $OA = 8$, the area of the shaded region $BAYB$ is

                  A. $16(\pi - 2)$ **D.** $64(\pi - 1)$

                  B. $16(\pi - 1)$ **E.** $32(\pi - 2)$

                  C. $8(2\pi - 1)$

Exercise 3:   If $CD = 10$, the perimeter of sector $DOCXD$ is

                  A. $30 + \dfrac{10\pi}{3}$ **D.** $20 + \dfrac{20\pi}{3}$

                  B. $30 + \dfrac{20\pi}{3}$ **E.** $30 + 30\pi$

                  C. $20 + \dfrac{10\pi}{3}$

Exercise 4:   If $OC = 2$, the area of the shaded portion $DCXD$ is

                  A. $\pi - \sqrt{3}$ **D.** $\dfrac{2\pi - 3\sqrt{3}}{3}$

                  B. $2\pi - \sqrt{3}$ **E.** $\dfrac{8\pi - 3\sqrt{3}}{3}$

                  C. $4\pi - \sqrt{3}$

Exercise 5:   If the area of $\triangle COD$ is $25\sqrt{3}$, the perimeter of semicircle $EOBDXCE$ is

                  A. 10π **D.** $10(2\pi + 1)$

                  B. $10(\pi + 1)$ **E.** $20(\pi + 1)$

                  C. $10(\pi + 2)$

I guess you get the idea already.

Let's look at the answers.

Answer 1:   C: $A = \frac{1}{2}(r)(r) = \frac{1}{2}r^2$, so $r^2 = 50$. The area of the circle is thus $\pi r^2 = 50\pi$. Notice that once we have a value for r^2, we don't have to find r to do this problem.

Answer 2:   A: The area of region *BAYB* is the area of one-fourth of a circle minus the area of $\triangle AOB$. So the area is $A = \frac{1}{4}\pi 8^2 - \frac{1}{2}(8)(8) = 16\pi - 32 = 16(\pi - 2)$.

Answer 3:   C: The perimeter of sector $DOCXD = 2r + s$, where s is the length of arc *CXD*. $\triangle COD$ is equilateral, so $CD = CO = DO = r = 10$. $s = \frac{60°}{360°}2\pi(10) = \frac{10\pi}{3}$. So the perimeter of sector *CODXD* is $p = 20 + \frac{10\pi}{3}$.

Answer 4:   D: The area of region *DCXD* is the area of sector *ODXCO* minus the area of $\triangle DOC$, when $OC = 2$. So the area is $A = \frac{60°}{360°}\pi 2^2 - \frac{2^2\sqrt{3}}{4} = \frac{2\pi}{3} - \sqrt{3} = \frac{2\pi - 3\sqrt{3}}{3}$.

Answer 5:   C: We must first find the radius. The area of equilateral $\triangle COD = \frac{s^2\sqrt{3}}{4} = 25\sqrt{3}$. So $s^2 = 100$, and $s = r = 10$. The perimeter of the semicircle is $\frac{1}{2}(2\pi r) + 2r = \pi r + 2r = 10\pi + 20 = 10(\pi + 2)$.

There are a few more things we need to know. When we talked about two intersecting lines earlier, we saw that *CE* might equal *ED*; however, if the description of the figure doesn't say so, you cannot assume it. Also, *ED* might be perpendicular to *CE*, but if it doesn't say so, you cannot assume it, either. In fact, we can say *CD* **bisects** *AB* at *E* only if we know that $AE = EB$ or *E* is the **midpoint** of *AB*.

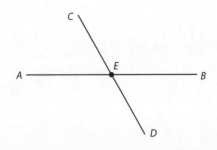

Now, however, we consider two intersecting lines in a circle, such as chord *AB* and radius *CO* in circle *O*.

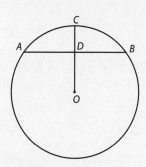

If one of the following facts is true, all are true:

1.   *OD* ⊥ *AB*

2.   *CDO* bisects *AB*

3.   *CO* bisects $\overset{\frown}{ACB}$ (read "arc *ACB*")

Q   **Let's do some more exercises**

For Exercises 6 and 7, use this figure, which is a triangle-semicircle shape.

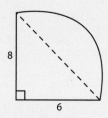

Exercise 6:   The perimeter of this figure is

A. $14 + 5\pi$ D. $24 + 10\pi$

B. $14 + 10\pi$ E. $12 + 10\pi$

C. $24 + 5\pi$

Exercise 7:   The area of this figure is

A. $24 + \dfrac{25\pi}{2}$ D. $48 + 25\pi$

B. $24 + 25\pi$ E. $48 + 50\pi$

C. $24 + 50\pi$

Exercise 8: A circle is **inscribed** in (inside and touching) figure *MNPQ*, which has all right angles. Diameter *AB* = 10. The area of the shaded portion is

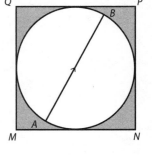

    A. 100 − 12.5π   **D.** 40 − 5π

    B. 100 − 25π   **E.** 100 − 100π

    C. 40 − 10π

Exercise 9: If *AB* = 10, the area of the shaded portion in the figure is

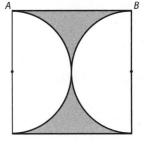

    A. 100 − 12.5π   **D.** 40 − 5π

    B. 100 − 25π   **E.** 100 − 100π

    C. 40 − 10π

For Exercises 10 and 11, use this figure. The perimeter of the 16 semicircles is 32π.

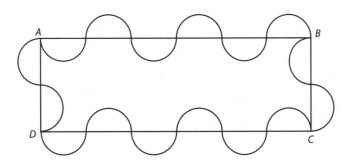

Exercise 10: The area of rectangle *ABCD* is

    A. 64   **D.** 16π

    B. 128   **E.** Cannot be determined

    C. 192

Exercise 11: The area inside the region formed by the semicircular curves from *A* to *B* to *C* to *D* and back to *A* is

    A. 64   **D.** 16π

    B. 128   **E.** Are you for real??!!

    C. 192

Exercise 12: In the figure, $EF = CD = 12$, B is the midpoint of OD, and A is the midpoint of CO. The area of the shaded portion is

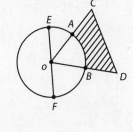

A. $36(4\sqrt{3} - \pi)$ D. $72(3\sqrt{3} - \pi)$

B. $6(6\sqrt{3} - \pi)$ E. $36(6\sqrt{3} - \pi)$

C. $144(\pi - 3)$

For Exercises 13 and 14, use this figure.

Exercise 13: If the diameter = 20, $OC \perp AB$, and $\angle A = 30°$, $AB =$

A. 10   D. $10\sqrt{3}$

B. 5   E. $5\sqrt{2}$

C. $5\sqrt{3}$

Exercise 14: If OC bisects AB, $AB = 16$, and $OC = 6$, the area of circle O is

A. 10π D. 100π

B. 20π E. 400π

C. 40π

 Let's look at the answers.

Answer 6:   A: The figure is a 6-8-10 Pythagorean triple, but 10 is not part of the perimeter. $p = 6 + 8 + \frac{1}{2}2\pi(5) = 14 + 5\pi$.

Answer 7:   A: $A = \frac{1}{2}bh + \frac{1}{2}\pi r^2 = \frac{1}{2}6 \times 8 + \frac{1}{2}\pi 5^2 = 24 + \frac{25\pi}{2}$. The answer is A.

Answer 8:   **B:** The area is the area of the square minus the area of the circle.
$A = s^2 - \pi r^2 = 10^2 - \pi 5^2 = 100 - 25\pi$.

Answer 9:   **B:** Answer 8 and Answer 9 are exactly the same problems. In Answer 8, we could also say the square **circumscribes** the circle.

Answer 10:   **C:** Each semicircle has arc length $\dfrac{180°}{360°}\pi d$, and there are 16 semicircles, so $16\left(\dfrac{1}{2}\pi d\right) = 32\pi$, or $8\pi d = 32\pi$, so $d = 4$. The rectangle's dimensions are thus 8 and 24. The area is $A = b \times h = 8(24) = 192$.

Answer 11:   **C:** Believe it or not, Exercise 11 is exactly the same as Exercise 10! We can think of the areas of the "outer" semicircles as canceling out the areas of the "inner" semicircles, and we are left with only the area of rectangle *ABCD*.

Answer 12:   **B:** The information is enough to tell us the triangle is equilateral and $\angle AOB = 60°$. The shaded area is the area of $\triangle COD$ minus the area of sector *OABO*. Thus, $A = \dfrac{s^2\sqrt{3}}{4} - \dfrac{1}{6}\pi r^2 = \dfrac{12^2\sqrt{3}}{4} - \dfrac{1}{6}\pi 6^2 = 36\sqrt{3} - 6\pi = 6(6\sqrt{3} - \pi)$.

Answer 13:   **D:** *AO*, the radius, is 10; *CO*, the side opposite the 30° angle, is 5; and *AC*, the side opposite the 60° angle, is $5\sqrt{3}$. $AB = 2(AC) = 2(5\sqrt{3}) = 10\sqrt{3}$.

Answer 14:   **D:** To find the area of the circle, we need to find the radius *OA*. We know *AC* is 8 and *OC* is 6. We have a 6-8-10 right triangle, so $AO = r = 10$. The area is $\pi(10)^2 = 100\pi$.

Okay. Now let's go from two dimensions to three dimensions.

CHAPTER 13: *Three-Dimensional Figures*

"Many dimensions in your trip will add to your ultimate success."

This chapter is relatively short. There are only a few figures we need to know. Since these are three-dimensional figures, we discuss their volumes and surface areas (areas of all of the sides). The diagonal is the distance from one corner internally to an opposite corner.

BOX

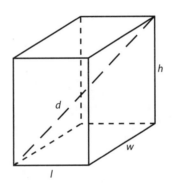

This figure is also known as a **rectangular solid**, and if that isn't a mouthful enough, its correct name is a **rectangular parallelepiped**. But essentially, it's a **box**.

- Volume $= V = \ell wh$

- Surface area $= SA = 2\ell w + 2\ell h + 2wh$

- Diagonal $= d = \sqrt{\ell^2 + w^2 + h^2}$, known as the 3-D Pythagorean Theorem

Example 1:   For the given figure, find V, SA, and d.

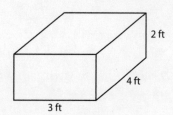

2 ft

4 ft

3 ft

Solution: $V = \ell wh = (3)(4)(2) = 24$ cubic feet; $SA = 2\ell w + 2\ell h + 2wh = 2(3)(4) + 2(3)(2) + 2(4)(2) = 52$ square feet; $d = \sqrt{\ell^2 + w^2 + h^2} = \sqrt{3^2 + 4^2 + 2^2} = \sqrt{29} \approx 5.4$ feet

CUBE

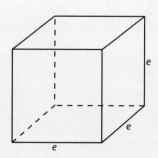

e

e

e

A **cube** is a box for which all of the faces, or sides, are equal squares.

- $V = e^3$ (read as "e cubed"). Cubing comes from a cube!
- $SA = 6e^2$
- $d = e\sqrt{3}$
- A cube has 6 faces, 8 vertices, and 12 edges.

Example 2:   For a cube with an edge of 10 meters, find V, SA, and d.

Solution: $V = 10^3 = 1,000$ cubic meters; $SA = 6e^2 = 6(10)^2 = 600$ square meters; $d = e\sqrt{3} = 10\sqrt{3} \approx 17.32$ meters

CYLINDER

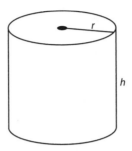

A cylinder is shaped like a can. The curved surface is considered as a side, and the top and bottom are equal circles.

- $V = \pi r^2 h$

- $SA = \text{top} + \text{bottom} + \text{curved surface} = 2\pi r^2 + 2\pi rh$

Once a neighbor of mine wanted to find the area of the curved part of a cylinder. He wasn't interested in why; just the answer. Of course, being a teacher I had to explain it to him. I told him that if he cut a label off a soup can and unwrapped it, the figure is a rectangle; neglecting the rim, the height is the height of the can and the width is the circumference of the circle. Multiply this height and width, and the answer is $2\pi r \times h$. He waited patiently and then soon moved. (Just kidding!)

In general, the volume of any figure for which the top is the same as the bottom is $V = Bh$, where B is the area of the base. If the figure comes to a point, the volume is $\left(\dfrac{1}{3}\right)Bh$. The surface area is found by adding up all the sides.

Example 3:   Find V and SA for a cylinder of height 10 yards and diameters of 8 yards.

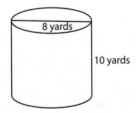

8 yards

10 yards

Solution: We see that since $d = 8$, $r = 4$. Then $V = \pi r^2 h = \pi(4^2 \times 10) = 160\pi$ cubic yards; $SA = 2\pi r^2 + 2\pi rh = 2\pi 4^2 + 2\pi(4)(10) = 112\pi$ square yards.

Q **Let's do some exercises.**

Use this figure for Exercises 1 through 3. It is a pyramid with a square base. $WX = 8$, $BV = 3$, and B is in the middle of the base.

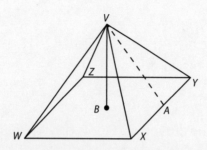

Exercise 1: The volume of the pyramid is

  A. 192   **D.** 32

  B. 96   **E.** 16

  C. 64

Exercise 2: The surface area of the pyramid is

  A. 72   **D.** 224

  B. 112   **E.** 448

  C. 144

Exercise 3: $VY =$

  A. 6   **D.** 9

  B. $\sqrt{41}$ **E.** $\sqrt{89}$

  C. 7

Exercise 4: In the given rectangular solid, the perimeter of $\triangle ABC =$

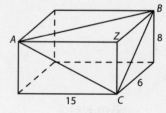

  A. $\sqrt{325} = 5\sqrt{13}$ **D.** 37

  B. 30   **E.** 41

  C. $27 + \sqrt{261}$

Exercise 5:   The volume of the cylinder shown is:

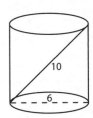

A. 640π D. 144π

B. 320π E. 72π

C. 288π

Exercise 6:   *ABKL* is the face of a cube with *AB* = 10, and box *BCFG* has a square front with *BC* = 6. The surface area that can be viewed in this configuration is

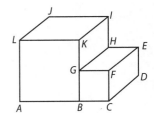

A. 300   D. 400

B. 356   E. 1360

C. 396

Exercise 7:   A cylinder has volume *V*. If we triple its radius, what do we multiply the height by in order for the volume to stay the same?

A. $\dfrac{1}{9}$ D. 3

B. $\dfrac{1}{3}$ E. 9

C. 1

 Let's look at the answers.

Answer 1:   C: $V = \left(\dfrac{1}{3}\right)Bh = \dfrac{1}{3}s^2h = \dfrac{1}{3}(8^2)(3) = 64.$

Answer 2:   C: $SA = s^2 + 4\left(\dfrac{1}{2}bh\right).$ $AB = \left(\dfrac{1}{2}\right)WX = 4;$ $\triangle ABV$ is a 3-4-5 right triangle with $AB = 4$ and $BV = 3$, so $AV = 5.$ *AV* is the height of each triangular side, *h*, and $b = XY = 8.$ So $SA = 8^2 + 2(8)(5) = 144.$

Answer 3:   B: $\triangle AVY$ is a right triangle with right angle at *A*. $AY = 4$ and $AV = 5$, so $VY = \sqrt{4^2 + 5^2} = \sqrt{41}.$

Answer 4:   C: In the given figure, we have to use the 2-D Pythagorean Theorem three times to find the sides of $\triangle ABC$. $\triangle BCY$ is a 6-8-10 triple, so $BC = 10$, and $\triangle ACX$ is a 8-15-17 triple, so $AC = 17$. For $\triangle ABZ$, we actually have to calculate the missing side $\sqrt{6^2 + 15^2} = \sqrt{261}$. So the perimeter is $10 + 17 + \sqrt{261}$.

Answer 5:   E: The diameter of the base is 6, and again we have a Pythagorean triple; so $h = 8$. The volume is $\pi(3^2)(8) = 72\pi$.

Answer 6:   C: The areas are: $ABKL = 100$; $IJLK = 100$; $BCEG = 36$; $CDEF = 60$; $EFGH = 60$; $GHIK = 40$. The total is 396.

Answer 7:   A: The volume $V = \pi r^2 h$. For simplicity, let $r = 1$ and $h = 1$. So $V = \pi$. If we triple the radius, $V = \pi(3)^2 h$. For the original volume to still be π, $9h = 1$, or $h = \dfrac{1}{9}$.

CHAPTER 14: *Miscellaneous Miscellany*

"At this point, your probability of success has greatly increased."

This chapter is a combination of related topics. Let's start with measures of central tendency.

CENTRAL TENDENCY

People say you can prove anything with statistics. You need to know what you are taking statistics of, how many in the group, and many other variables. Even then, depending on the "spin" you want, you can choose among three measures to describe the data and prove your point. These measures of central tendency are ways to find a "typical" value. The measures of central tendency, introduced in Chapter 1, are:

Mean: Add up the number of terms and divide the sum by the number of terms; that's the way grades are usually determined in school.

Median: The middle term when the numbers are put in order from smallest to largest (or the other way around); for an odd number of terms, it is the middle term; for an even number, it's the mean of the middle two numbers.

Mode: The most common term; there can be one mode, two modes (bimodal), or any number of modes.

Example 1: For the data consisting of the numbers 5, 6, 8, 9, 12, 12, 18, what is the mean, median, and mode?

Solution: The mean is $\dfrac{5 + 6 + 8 + 9 + 12 + 12 + 18}{7} = \dfrac{70}{7} = 10$. The median is 9. There are three numbers above it and three numbers below it.

The mode is 12. It is the most common number, appearing twice.

Now can you see how you can prove anything with statistics?

Q   **Let's do some exercises.**

For Exercises 1–3, use the following numbers: 8, 10, 10, 16, 16, 18

Exercise 1:   The mean is

A. 8   D. 16

B. 10   E. There are two of them

C. 13

Exercise 2:   The median is

A. 8   D. 16

B. 10   E. There are two of them

C. 13

Exercise 3:   The mode is

A. 8   D. 16

B. 10   E. There are two of them

C. 13

   Let's look at the answers.

Answer 1:   C: $\dfrac{8 + 10 + 10 + 16 + 16 + 18}{6} = 13$.

Answer 2:   C: There are an even number of numbers, so we have to take the average of the middle two: $\dfrac{10 + 16}{2} = 13$.

Answer 3:   E: It's bimodal; the modes are 10 and 16, each appearing twice.

Sometimes statistics are given in frequency distribution tables, such as this one showing the grades Sandy received in 10 English quizzes.

Sandy's Quiz Scores

Grade	Number
100	4
98	3
95	2
86	1
Total	10

Example 2:   Find the measures of central tendency of Sandy's quiz scores.

Solution: The mean is the longest measure to compute:
$\dfrac{4 \times 100 + 3 \times 98 + 2 \times 9 + 86}{10} = 97$. The median is determined by
putting all of the numbers in order, so we have 100, 100, 100, 100, 98, 98, 98, 95, 95, 86. The middle terms are 98 and 98, so the median is 98. The mode is 100 since that is the most common score; there are four of them.

STANDARD DEVIATION

Look at these two sets of numbers:

Set *A*: {16, 18, 18, 19, 22, 23, 24}

Set *B*: {2, 18, 18, 19, 25, 28, 30}

If we find the median for each set, it is 19; if we find the mean for each set, it is 20; if we find the mode for each set, it is 18. The measures of central tendency are the same for both sets. However, there is something different about each set. In set *A*, all the numbers are relatively close to the mean. In set *B*, that is not true. We can measure the spread of the data by finding the **standard deviation**. Here are the steps to calculate it:

1. Find the mean of the set.

2. For each number, subtract the mean and square the result.

3. Add these squares together and divide by the number of elements in the set.

4. Take the square root of this result. That is the standard deviation.

The standard deviation of set *A* is:

$$\sqrt{\dfrac{(16-20)^2 + (18-20)^2 + (18-20)^2 + (19-20)^2 + (22-20)^2 + (23-20)^2 + (24-20)^2}{7}} \approx 2.78$$

The standard deviation of set B is:

$$\sqrt{\frac{(2-20)^2 + (18-20)^2 + (18-20)^2 + (19-20)^2 + (25-20)^2 + (28-20)^2 + (30-20)^2}{7}}$$

≈ 8.64

By comparing these two results, we have proof that the numbers in set B are more spread out, or dispersed, than those in set A. The standard deviation is a useful measure, especially in relation to the normal or bell-shaped curve. For example, by using multiples of the standard deviation (there are published tables for this), a manufacturer can determine how many items to produce. Suppose the manufacturer wanted to produce 100,000 pairs of a particular shoe. The standard deviation will tell how many of each size to produce. It also says that if you are a man wearing size 15 or a woman wearing size 12, you must go to a specialty store. If you are a man wearing size 5 or a woman wearing size 3, most of your shoes are children's shoes. The statistics tell you that it doesn't pay to make many shoes, if any, in those sizes.

This topic will be on the GMAT exam. In all honesty, they couldn't ask the question this way because you would need a calculator for almost every example. However, you should be familiar with the vocabulary presented here.

COUNTING

The **basic law of counting** says: "If you can do something in p ways, and a second thing in q ways, and a third thing in r ways, and so on, the total number of ways you can do the first thing, then the second thing, then the third thing , etc., is $p \times q \times r \times \ldots$

> **Example 3:**   If we have a lunch choice of 5 sandwiches, 4 desserts, and 3 drinks, and we can have one of each, how many different meals could we choose?
>
> **Solution:**   We can choose from $(5)(4)(3) = 60$ different meals

Arrangements

Let $n(A)$ be the number of elements in set A. In how many ways can these elements be arranged? The answer is that the first has n choices, the second has $(n - 1)$ choices (since one is already used), the third has $(n - 2)$ choices, all the way down to the last element, which has only one choice. In general, if there are n choices, the number of ways to choose is $n!$ (read as "**n factorial**") $= n(n - 1)(n - 2) \times \ldots (3)(2)(1)$.

> **Example 4:**   How many ways can five people line up?
>
> **Solution:**   This is just $5 \times 4 \times 3 \times 2 \times 1 = 120$.

Example 5: How many ways can 5 people sit in a circle?

Solution: It would appear to be the same question as Example 4, but it's not. If we draw the picture, each of 5 positions would be the same. The answer is (5)(4)(3)(2)(1) ÷ 5 = (4)(3)(2)(1) = 24. So n people can sit in a circle in $(n − 1)!$ ways.

Permutations

Permutations are essentially the law of counting without repeating, but order is important.

Example 6: How many ways can 7 people occupy 3 seats on a bench?

Solution: Any one of 7 people can be in the first seat, then any one of 6 people can be in the second seat, and any one of 5 people can be in the third seat. The total number would be (7)(6)(5) = 210 ways. There are many notations for permutations. One notation for this example would be $P(7, 3)$.

Combinations

Combinations are essentially the law of counting, with no repetition, and order doesn't matter.

Example 7: How many sets of three different letters can be made from eight different letters?

Solution: Since order doesn't matter, unlike with permutations, AB is the same as BA. So we can take the number of permutations, but we have to divide by the number of duplicates. It turns out that the duplicates for 3 letters is $3 \times 2 \times 1 = 6$. So we would have $\dfrac{8 \times 7 \times 6}{3 \times 2 \times 1} = 56$.

Again, there are many notations for combinations. One notation for this example is $C(8, 3)$.

Avoiding Duplicates

When we count how many ways to do A or B, we should be careful not to count any item twice. We must subtract out any items that include both A and B:

$$N(A \text{ or } B) = N(A) + N(B) − N(A \text{ and } B)$$

Example 8: Thirty students take French or German. If 20 took French and 18 took German, and if each student took at least one language, how many took both French and German?

Solution: $N(A \text{ or } B) = N(A) + N(B) - N(A \text{ and } B)$ or $30 = 20 + 18 - x$, so $x = 8$ took both languages.

Example 9: Forty students take Chinese or Japanese. If 9 take both and 20 take Japanese, how many students take Chinese?

Solution: $N(C \text{ or } J) = N(C) + N(J) - N(\text{both})$, or $40 = x + 20 - 9$, so $x = 29$ take Chinese.

Ⓠ **Let's do some exercises.**

For Exercises 4–8, use the set $\{e, f, g, h, i\}$. A word is considered to be any group of letters together; for example, hhg is a three-letter word.

Exercise 4: From this set, the number of three-letter words is:

A. 6   D. 60

B. 27   E. 125

C. 30

Exercise 5: How many three-letter permutations are there in this set?

A. 6   D. 60

B. 27   E. 125

C. 30

Exercise 6: How many three-letter words starting with a vowel and ending in a consonant can be made from this set?

A. 6   D. 60

B. 27   E. 125

C. 30

Exercise 7:   How many three-letter words with the second and third letters the same can be made from this set?

A. 5   D. 60

B. 20   E. 125

C. 25

Exercise 8:   How many three-letter permutations with the first and last letters *not* vowels can be made from this set?

A. 18   D. 45

B. 27   E. 125

C. 30

Exercise 9:   Fifty students take Spanish or Portuguese. If 20 take both and 40 take Spanish, the number of students taking Portuguese *only* is

A. 0   D. 20

B. 5   E. 30

C. 10

 Let's look at the answers.

Answer 4:   E: (5)(5)(5) = 125.

Answer 5:   D: (5)(4)(3) = 60.

Answer 6:   C: The first letter has 2 choices, the second can be any (5), and the third has 3 choices, so (2)(5)(3) = 30.

Answer 7:   C: There are 5 choices for the first two letters, but there is only 1 choice for the third letter since it must be the same as the second, so (5)(5)(1) = 25.

Answer 8:   A: There are 3 choices for the first letter, but only 2 choices for the last letter since it can't be a vowel and must be different than the first letter. There are three choices for the middle letter since two letters have already been used; so the answer is (3)(3)(2) = 18. These questions must be read very carefully!

Answer 9: **C:** This is not quite the same as the previous exercise. $N(S \text{ or } P) = N(S) + N(P) - N(\text{both})$; $50 = 40 + x - 20$; $x = 30$, but that is not the answer. If 30 take Portuguese and 20 take both, then 10 take Portuguese only.

PROBABILITY

The probability of an event is the number of "good" outcomes divided by the total number of outcomes possible, or $Pr(\text{success}) = \dfrac{\text{good outcomes}}{\text{total outcomes}}$.

Example 10: Consider the following sets: {26-letter English alphabet}; vowels = {a, e, i, o, u}; consonants = {the rest of the letters}. What are the probabilities of choosing a vowel? a consonant? any letter? π?

Solution: $Pr(\text{vowel}) = \dfrac{5}{26}$; $Pr(\text{consonant}) = \dfrac{21}{26}$; $Pr(\text{letter}) = \dfrac{26}{26} = 1$; $Pr(\pi) = \dfrac{0}{26} = 0$.

Probability follows the same rule about avoiding duplicates as discussed in the previous section.

$$Pr(A \text{ or } B) = Pr(A) + Pr(B) - Pr(A \text{ and } B)$$

Example 11: What is the probability that a spade or an ace is pulled from a 52-card deck?

Solution: $Pr(\text{Spade or ace}) = Pr(\text{Spade}) + Pr(\text{Ace}) - Pr(\text{Spade ace}) =$
$\dfrac{13}{52} + \dfrac{4}{52} - \dfrac{1}{52} = \dfrac{16}{52} = \dfrac{4}{13}$.

As weird as it sounds, whenever I taught this in a class, I never failed to have at least two students who didn't know what a deck of cards was, and I taught in New York City!

Use this figure for Examples 12 and 13. In the jar are 5 red balls and 3 yellow balls.

Example 12: What is the probability that two yellow balls are picked, with replacement?

Solution: $Pr(\text{2 yellow balls, with replacement}) = \left(\dfrac{3}{8}\right)\left(\dfrac{3}{8}\right) = \dfrac{9}{64}$

Example 13: What is the probability of picking two yellow balls, without replacement?

Solution: $Pr(\text{2 yellow balls, no replacement}) = \left(\dfrac{3}{8}\right)\left(\dfrac{2}{7}\right) = \dfrac{3}{28}$

CHARTS AND GRAPHS

The arithmetic on the actual GMAT could be less or more, nicer or messier, than the following exercises.

 Let's do some exercises.

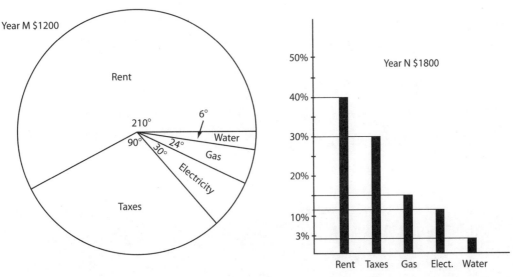

Some of the major expenses of the apartment of Mr. and Mrs. Smith in Smallville, USA, are shown in this pie chart and bar graph. The pie chart is for year *M* with a $1200 budget, and the bar graph is for year *N*, some years later, with an $1800 budget. Use these data for Exercises 10 through 14.

Exercise 10: The smallest percentage increase from year *M* to year *N* is for

    A. Rent   **D.** Gas

    B. Taxes   **E.** Water

    C. Electricity

Exercise 11: The largest percentage increase from year M to year N is for

    A. Rent   **D.** Gas

    B. Taxes   **E.** Water

    C. Electricity

Exercise 12: The change in rent from year M to year N was

    A. −$80   **D.** +$100

    B. none   **E.** +$190

    C. +$20

Exercise 13: The two closest monetary amounts are

    A. Rent in year M and   **D.** Taxes in year M and Gas in year N
        the Rent in year N

    B. Electricity in year M and   **E.** Electricity in year M and Electricity
        Gas in year N in year N

    C. Water in year M and
        Water in year N

Exercise 14: Which expenses exceeded the percentage increase in the total budget?

    A. All the expenses   **D.** All except Water and Rent

    B. All except Rent   **E.** All except Rent and Taxes

    C. All except Water

These exercises are easier to answer if we exactly calculate all of the money answers and put the items next to each other in a table:

	Year M	Year N
Rent	$\frac{210}{360} \times \$1200 = \frac{7}{12} \times \$1200 = \$700$	$.40 \times \$1800 = \$720$
Taxes	$300	$540
Electricity	$100	$216
Gas	$80	$270
Water	$20	$54

Let's look at the answers.

Answer 10:   **A:** Rent increased by only $20 (due perhaps to rent control or family member owner); the percentage increase is the smallest increase $\left(= \dfrac{20}{700} \times 100 \right)$. We don't actually have to calculate the exact percentage. We only have to note the percentage increase is obviously much smaller than the percentage increase of any other item.

Answer 11:   **D:** The percentage increase for gas is $\dfrac{190}{80} \times 100\%$, or more than a 200% increase.

Answer 12:   **C:** $720 − $700 = $20.

Answer 13:   **A:** The rents in year M and year N are only $20 apart. No other choices are this close.

Answer 14:   **B:** The total increase from year M to year N is 50%; taxes almost doubled; electricity more than doubled; gas more than tripled, and water almost tripled.

The GMAT doesn't usually ask more than one question on graphs or charts.

SETS

We have mentioned the topic of sets informally. Now let's be a little more formal.

Sets are denoted by braces $\{a, b, c\}$. This set has three elements: a, b, and c.

Example 14: $A = \{a, b\}$, $B = \{b, a\}$, $C = \{a, b, b, b, a, b, b, a, a\}$. How is set A related to set B? To set C?

Solutions: $A = B$, since the order does not matter in sets. $A = C$, since repeated elements are counted only once. C has only two elements in it: a and b.

We write $A \cup B$, read "A **union** B." as the set of elements in A or in B or in both.

We write $A \cap B$, read "A **intersect(ion)** B." as the set of elements common to A and B.

The **null set**, written $\{\}$ or ϕ (the Greek letter phi) is the set with nothing in it.

Example 15: Let $D = \{a, b, c, d, e, f\}$, $E = \{c, d, f, g\}$, $F = \{a, b, e\}$. Find $D \cup E$, $D \cap E$, and $E \cap F$.

Solution: $D \cup E = \{a, b, c, d, e, f, g\}$; $D \cap E = \{c, d, f\}$; $E \cap F = \phi$. Such sets are called **disjoint**.

We say A is a subset of B, written $A \subseteq B$ if every element in A is also in B. If a set has n elements, it has 2^n subsets.

Example 16: Write all the subsets of $G = \{a, b, c\}$.

Solution: Since G has 3 elements, there are $2^3 = 8$ subsets. They are: ϕ, $\{a\}$, $\{b\}$, $\{c\}$, $\{a, b\}$, $\{a, c\}$, $\{b, c\}$, $\{a, b, c\}$

Two other terms relating to sets that we need to know are:

- The **universe**, which is the set of everything we are talking about.

  Example 17: The universe U could be {all animals}, or it could be A = {all mammals.} It depends on what our topic of interest is. Note that $A \subseteq U$.

- A **complement** is the set of elements in the universe not in a specific set. The complement is denoted by a superscript letter c, so A^c is the complement of A. There are many other symbols in math for the complement.

  Example 18: If U = {all animals} and M = {all mammals}, then M^c = {reptiles, insects, fish, etc.}

  Example 19: If the alphabet is the universe, and A = {vowels}, then A^c = {consonants}.

 If we change the universe, we also automatically change the complement. This type of complement is spelled with two e's; not as "compliment," which is flattering.

EXPONENTIAL GROWTH AND DECAY

This last topic in the book is perhaps the most fascinating to discuss in a nonmathematical way. This topic relates to population growth, earthquakes, the age of dinosaurs, and the real destructive properties of an atomic bomb.

Let's mention a little about the first one. Assume human beings first appeared on Earth 40,000 years ago. If we counted every human being who lived from that time until the year 1900 (probably 1920 by now), there are still more people alive today than that total number. That is an example of exponential growth.

Let's see how we do such problems.

Example 20:   Deb started with $10,000. If her net worth triples every five years, how much will she have in 30 years?

Solution: In 5 years, the $10,000 will be $30,000; in 10 years, $90,000; in 15 years $270,000; in 20 years, $810,000; in 25 years, $2,430,000; and in 30 years, $7,290,000!

Example 21:   Radioactive goo has a half-life of 15 minutes. If there are 240 pounds of radioactive goo, how much radioactive goo is left in an hour? (A half-life is the time required for half of a radioactive substance to disintegrate.)

Solution: In 15 minutes, the 240 pounds will be 120 pounds; in 30 minutes, that 120 pounds will be 60 pounds; in 45 minutes, there will be 30 pounds; and in one hour, only 15 pounds will be left.

Now let's try problems—lots of problems.

CHAPTER 15: *Data Sufficiency Questions*

"If your data is sufficient, you will do well on this test."

Before we do more problem-solving, we will take a look at a different kind of problem: data sufficiency.

You will be given a problem and two statements. You must decide whether the information in the statements is sufficient to solve the problem. You do *not* have to solve the problems, just determine if they can be solved with the given information. Your response on each data sufficiency question will be one of these choices, depending on whether the information given in the statements is sufficient to answer the question:

A. Statement (1) ALONE is sufficient, but statement (2) alone is not sufficient.

B. Statement (2) ALONE is sufficient, but statement (1) alone is not sufficient.

C. BOTH statements TOGETHER are sufficient, but NEITHER statement ALONE is sufficient.

D. EACH statement ALONE is sufficient.

E. Statements (1) and (2) TOGETHER are NOT sufficient.

At this point, you may want to look at the examples. However, let us first talk about how to attack these problems.

- Look at statement (1). If it is sufficient, the answer must be A or D. Then look at statement (2). If statement (2) is true, the answer is D. If statement 2 is false the answer is A.

- Suppose statement (1) is not sufficient; then the answer must be B, C, or E. Then look at statement (2). If statement 2 is sufficient, the answer is B. If statement (2) is not sufficient, but statements (1) and (2) taken together are sufficient, the answer is C. If statements (1) and (2) are both not sufficient and together they also are not sufficient, the answer is E.

- Remember: You do not have to solve the problems. The solutions presented in this chapter may solve the problems, but that is for information purposes only; the solutions are not required for the test. You are asked only to determine *whether* the problem can be solved with the given information.

Whew! Let's give a simple example of each.

Example 1: Find the volume of the cube.

   (1) The edge of the cube is 5.

   (2) The cube is made of maple.

Solution: A: $V = e^3$. Statement 1 gives the edge; the information is sufficient. Statement 2 is not sufficient because it says nothing of the length of the edge.

Example 2: Find the volume of a child's ball.

   (1) The ball is pink.

   (2) The radius of the ball is 2 inches.

Solution: B: $V = \frac{4}{3}\pi r^3$. Statement 1 is not sufficient, but statement 2 is sufficient.

Example 3: The area of the rectangle.

   (1) The length is 5 feet.

   (2) The width is 3 meters.

Solution: C: Area = length times width. Neither statement by itself is sufficient, but together, it is enough information to solve the problem. It might be a messy conversion, but you do not have to actually get the area. You only need to know that you can get the area with both statements 1 and 2 together.

Example 4: The area of a circle.

   (1) The radius is 10 inches.

   (2) The diameter is 20 inches.

Solution: D: $A = \pi r^2$. Statement 1 is sufficient since $r = 10$. Statement 2 is sufficient since if $d = 20$, $r = \frac{20}{2} = 10$.

Example 5: The volume of a box.

      (1) The length is 5 feet.

      (2) The height is 6 feet.

Solution: E: $V = \ell \times w \times h$. Since three dimensions are needed and the two statements together give only two dimensions, we have insufficient information to solve the problem.

Let's do lots of problems.

Answer Sheets for Tests A-D are at the back of the book.

Ⓠ Data Sufficiency Practice Test A

  A. Statement (1) ALONE is sufficient, but statement (2) alone is not sufficient.

  B. Statement (2) ALONE is sufficient, but statement (1) alone is not sufficient.

  C. BOTH statements TOGETHER are sufficient, but NEITHER statement ALONE is sufficient.

  D. EACH statement ALONE is sufficient.

  E. Statements (1) and (2) TOGETHER are NOT sufficient.

A1: The volume of a given cylinder.

    (1) $r^2h = 7$.

    (2) The radius is half the height.

A2: Is x an even or odd integer?

    (1) x is an integer and x^2 is odd.

    (2) x is an integer and x^2 is even.

A3: The area of a ring.

    (1) The difference in the radii is 2.

    (2) The inside radius is 7.

A4: The equation of the parabola $y = ax^2 + bx + c$.

    (1) The parabola has x-intercepts (7, 0) and (1, 0) and vertex (4, −18).

    (2) The parabola passes through the points (2, −10), (3, −16), and (6, −10).

A5:   The surface area of a cube.

    (1)   The volume is 125.

    (2)   The diagonal is $5\sqrt{3}$.

A6:   Does $x = y$?

    (1) $x^3 = y^3$.

    (2) $x^4 = y^4$.

A7:   Is $ab < 10$:

    (1) $a < 2$ and $b < 5$.

    (2) $a^2 < 36$ and $\dfrac{1}{3} < b < \dfrac{2}{3}$.

A8:   Is $a < b < c$?

    (1) $ab < bc$.

    (2) $ab < ac$.

A9:   The area of the rectangle:

    (1)   A triangle with the same base and same height as the rectangle has an area of 45 square feet.

    (2)   A parallelogram with the same base and same height as the rectangle has an area of 90 square feet.

A10:   In the line $Ax + By = C$, determining A and B:

    (1)   Slope 2 and y-intercept $(0, -8)$.

    (2) x-intercept $(4, 0)$ passing through the point $(3, -2)$.

A11: m and n are positive integers. Is $m + n$ divisible by 3?

    (1) m is divisible by 3.

    (2) n is divisible by 6.

A12:   The value of x:

    (1) $\dfrac{1}{4x^2} = 1$.

    (2) $3x - 4 = 5x - 5$.

A13: Volume of a cylinder with height 8.

    (1) Circumference of the base is 6π

    (2) Diagonal of length 10 is drawn from the edge of one base through the center of the cylinder to the edge of the other base. The bases are circular and parallel to each other.

A14: The total amount of money earned:

    (1) Principal of $500, 4% interest for 2 years.

    (2) $40 interest.

A15: Determine three consecutive positive even integers a, b, and c:

    (1) b is the mean of a and c.

    (2) $a^2 + b^2 = c^2$.

A16: Is $x + y - z$ is greater than $x - y + z$?

    (1) $y < 0$.

    (2) $z > 0$.

A17: What is the value of the sum of p even integers:

    (1) $p = 10$.

    (2) $p^3 = 1{,}000$.

A18: Determine the value of x if $\dfrac{x^2}{y} = z$.

    (1) $yz = 36$.

    (2) $y = 72$; $z = \dfrac{1}{2}$.

A19: Determine who received the bigger raise.

    (1) Bea received a 7% raise.

    (2) Mel received a 5% raise.

A20: Determine the combined cost of a ball and a bat.

    (1) Three balls and 3 bats cost $42.

    (2) Five balls and 7 bats cost $90.

A21: Is p an odd number?

  (1) p is divisible by 11.

  (2) 7 is a factor of p.

A22: Determine the value of m if $2m + n + 7 = 4m - 5n + 9$.

  (1) $n^2 = 16$.

  (2) $n = 4$.

A23: Is $\dfrac{x}{y} > 1$?

  (1) $x - y > 0$.

  (2) $xy > 1$.

A24: Given s, t, u are integers; are $s, t,$ and u consecutive integers?

  (1) $s < t < u$.

  (2) $u = s + 2$.

A25: Determining the area of a rectangle with base b and height h:

  (1) $b + 2h = 20$.

  (2) $b + h = 16$.

A26: Is h negative?

  (1) h^5 is negative.

  (2) $-h$ is positive.

A27: If m is a positive integer, is $\sqrt{m}$ an integer?

  (1) m is the square of an integer.

  (2) $\sqrt{m}$ is the square of an integer.

A28: If $mn < 4$, is $m \le 1$?

  (1) $n > 4$.

  (2) $m < 2$.

A29: The value of $m^2 - n^2$.

  (1) $m + n = m - n$.

  (2) $m + n = \dfrac{1}{m - n}$.

A30: If x is a positive integer, is x odd?

   (1) $5x$ is odd.

   (2) $x + 17$ is even.

A31: In Isosceles $\triangle ABC$, A and C are the base angles. Can we find all the angles of the triangle?

   (1)   The sum of the angles of the triangle is $180°$.

   (2) $B = A + 30$.

A32: m, n are integers; the value of m.

   (1) $mn = 64$.

   (2) $m = n^2$.

A33: If $n = 0.abcd$, where a, b, c, and d are nonzero digits, find n.

   (1) $a = 2b = 4c = 8d$.

   (2) $ad = bc$.

A34: The value of $q + r$:

   (1) $pq + rs + pr + qs = 20$.

   (2) $p + s = 5$.

A35: n is an integer; is 9 a factor of n^2?

   (1) n is divisible by 6.

   (2)   12 is a factor of n.

A36: If r is an integer, is $\dfrac{120 - r}{r}$ an integer?

   (1) $r^2 = 36$.

   (2) $0 < r < 7$.

A37: Is $mx = 7 - nx$?

   (1) $x(m + n) = 7$.

   (2) $m = n = 2$ and $x = 1.75$.

Let's look at the answers.

A1: **A:** $V = \pi r^2 h$. Statement 1 is sufficient. Even though we don't specifically know r and h, we do know $r^2 h$, enough to find the volume. Statement 2 gives us the volume in terms of one variable but not the value of that variable, so not enough information to find the volume.

A2: **D:** Both of these conditions determine oddness or evenness. You must know these two facts: If x is an odd integer, x^2 is odd; if x is an even integer, x^2 is even.

A3: **C:** $A = \pi r^2_{\text{outer}} - \pi r^2_{\text{inner}}$. Statement 1 is not sufficient since we know only the difference in the radii. Statement 2 is insufficient because we have only the inside radius. Together, though, we also know the outside radius, 9, and can solve the problem. Remember you do *not* solve the problem; you only have to know that it *can* be solved.

A4: **D:** In giving three points in each case, we get three equations in three unknowns, which are solvable. By the way the parabola is $y = 2x^2 - 16x + 14$.

A5: **D:** $SA = 6e^2$. Statement 1 tells us $V = e^3 = 125$, so $e = 5$. Statement 2 tells us $d = e\sqrt{3} = 5\sqrt{3}$ and again $e = 5$.

A6: **A:** Statement 1 is sufficient; but statement 2 is not sufficient since $y = \pm x$.

A7: **B:** Statement 1 is not sufficient and very tricky; suppose $a = -5$ and $b = -6$, then $ab = 30$. Statement 2 is sufficient since $-6 < a < 6$; if you multiply it by any number between $\frac{1}{3}$ and $\frac{2}{3}$, it is still less than 10.

A8: **E:** Statement 1 is not sufficient; for example, if $a = 2, b = 3, c = 5$, then $ab < bc$ and $a < b < c$; however, if $a = 5, b = -3, c = 2$, then $ab < ac$ and $a > c > b$. Statement 2 is not sufficient. If $a = 2, b = 3, c = 5$, then $ab < ac$ and $a < b < c$; however, if $a = -2, b = 5, c = 3$, then $ab < ac$ and $a < c < b$. Together, statements 1 and 2 are not sufficient.

A9: **D:** Both sufficient; once you know the base and the height, you know the area of the rectangle since the triangle is half the rectangle, and the parallelogram has exactly the same area as the rectangle.

A10: **D:** Both statements are sufficient. Two points or the slope and one point uniquely determine a line.

A11: **C:** Statement 1 is not sufficient since 3 divides into m, but we don't know about n. In a sum, if 3 divides evenly into one term, it must evenly divide into both terms. Similarly statement 2 is not sufficient since 3 divides evenly into n (since 6 divides n), but we don't know about m. However, combining the statements gives 3 dividing evenly into both m and n.

A12: **B:** Statement 1 is not sufficient since $x = \pm\frac{1}{2}$. Statement 2 is sufficient since $x = \frac{1}{2}$.

A13: **D:** $V = \pi r^2 h$. You need to know r. Statement 1 is sufficient since the circumference $c = 2\pi r$; so $r = 3$. Statement 2 is sufficient because the diameter, height, and diagonal form a 6-8-10 right triangle, and the diameter is 6; so the radius is 3.

A14: **A:** Statement 1 is sufficient since $A = p + prt$, and p, r, t are all given. Statement 2 is not sufficient since the principal is not known.

A15: **B:** Statement 1 is not sufficient since this statement is true for all arithmetic sums. Statement 2 is sufficient; if you write $a^2 + b^2 = c^2$ in terms of the same variable and solve, you get $n^2 + (n + 2)^2 = (n + 4)^2$. Solving, you get the 6-8-10 Pythagorean triple.

A16: **C:** Statements 1 and 2 are not sufficient since we need to know about the other letter. Together, the statements show $x - y + z$ is greater.

A17: **E:** Neither statement individually or together gives you the sum because we don't know the first integer.

A18: **E:** Neither statement is sufficient since both give $x^2 = 36$; so $x = \pm 6$.

A19: **E:** Neither statement gives you a salary. So we know nothing about what either raise is in dollars.

A20: **A:** Statement 1 is sufficient since if 3 balls and 3 bats = \$42, then 1 ball and 1 bat = \$14. Statement 2 is not sufficient since 5 balls and 7 bats will not give you the price of one of each.

A21: **E:** Statement 1 is not sufficient because p could be 22 or 33. Statement 2 is not sufficient because p could be 7 or 14. Together the statements are not sufficient because p could be 77 or 154.

A22: **B:** Statement 1 is not sufficient since n has two values, ± 4, so m would have two possible values. Statement 2 is sufficient since $n = 4$ means $m = 11$.

A23: **E:** Statement 1 is not sufficient; look at $x = 4, y = 3$; and $x = -3, y = -4$. Statement 2 is not sufficient; look at $x = 5, y = 4$ and $x = 4, y = 5$. Together, the statements still are not sufficient.

A24: **C:** Statement 1 is not sufficient; look at $4 < 8 < 11$. Statement 2 is not sufficient, for example, $s = 5, t = 10, u = 7$. Both statements together are sufficient, since if $u = s + 2$ and $s < t < u$, then $t = s + 1$. This implies that s, t, u are consecutive integers.

A25: **C:** $A = bh$. Alone, each is not sufficient to determine both b and h; together, however, by subtracting the equations, we get $h = 4$ and $b = 12$. So the area is 48.

A26: **D:** Both statements are true independently to show h is negative.

A27: **D:** Statement 1 is sufficient: $m = a^2$, where a is an integer; so $\sqrt{m} = \pm a$, both integers. Statement 2 is sufficient since the square of an integer is an integer.

A28: **A:** Statement 1 is sufficient. If $n > 4, m < 1$. Statement 2 is not sufficient; m can be 1.5.

A29: **B:** Statement 1 is not sufficient; we get $n = 0$ but not the value of m. Statement 2 is sufficient; cross multiplying gives $m^2 - n^2 = 1$.

A30: **D:** Both are sufficient; odd $\times$ odd = odd; odd + odd = even.

A31: **B:** Statement 1 is not sufficient; this is true for all triangles. Statement 2 is sufficient; we can solve $A + A + (A + 30) = 180$ to get all the angles.

A32: **C:** Each statement is not sufficient since we can't find both m and n with only the individual statements. Together, however, by substituting, we get $n^3 = 64, n = 4$, $m = 16$.

A33: **A:** Statement 1 is sufficient; since these are all digits 0 through 9, a must equal 8; so $b = 4; c = 2; d = 1$. Statement 2 is not sufficient since $(1)(4)=(2)(2), (1)(6)= 2(3)$, etc., all can be true.

A34: **C:** Statements 1 and 2 are not sufficient; however, together they are $pq + pr + qs + rs = p(q + r) + s(q + r) = (p + s)(q + r) = 20; p + s = 5$; so $q + r = 4$.

A35: **D:** Both statements are sufficient; since 6 and 12 are factors of n, 3 must be a factor of n; so $(3)(3)$ is a factor of $n(n) = n^2$.

A36: **D:** Both statements are sufficient. For statement 1, substituting $+6$ or -6 gives an integer; statement 2 is sufficient because substituting 1, 2, 3, 4, 5, and 6 gives integers.

A37: **D:** Both are sufficient for entirely different reasons. By bringing nx to the other side of the equation and factoring, we get statement 1. By substitution, statement 2 is correct.

ⓠ Data Sufficiency Practice Test B

A. Statement (1) ALONE is sufficient, but statement (2) alone is not sufficient.

B. Statement (2) ALONE is sufficient, but statement (1) alone is not sufficient.

C. BOTH statements TOGETHER are sufficient, but NEITHER statement ALONE is sufficient.

D. EACH statement ALONE is sufficient.

E. Statements (1) and (2) TOGETHER are NOT sufficient.

B1: $a > b$, and $b > c$. Is $a > d$?

   (1) $c < d$.

   (2) $c > d$.

B2: The value of x and the value of y;

   (1) $2^x 8^y = 32$.

   (2) $3^x 9^y = 81$.

B3: How many hours to finish the job?

   (1) Sid takes 6 hours to do a job.

   (2) Cyd takes 5 hours to finish the same job.

B4: Find the edge of a cube.

   (1) The volume equals the surface area.

   (2) The ratio of the diagonal of the cube to the diagonal in the face of the cube is $\sqrt{3} : \sqrt{2}$.

B5: We define a right identity as the element b such that $a \# b = a$ for all elements a in the set; $a \# b = ab + a + b$. Does the right identity element exist in set S?

   (1) 0 is an element of S

   (2) 1 is an element of S

B6: Is line m perpendicular to line p?

   (1) $m \parallel n$, and $n \perp p$.

   (2) $m \perp n$ and $n \perp p$.

B7: BD bisects $\angle ABC$ and BE bisects $\angle DBC$. The number of degrees in $\angle EBC$.

   (1) $\angle ABC = 72°$.

   (2) $\angle ABD = 36°$.

B8: a, 7, and c form an arithmetic progression; find a and c.

   (1) $c = 5a$.

   (2) $\dfrac{a+c}{2} = 7$.

B9:   The value of n.

   (1) $(2^n)^n = 4^2$.

   (2) $([2^n]^n)^n = 4^4$.

B10:   Do we always get the original price?

   (1)   The price increases by 25% and the decreases by 20%.

   (2) $40 increase followed by a 10% decrease.

B11:   The distance between city A and city B is 20 miles, and the distance between city B and city C is 15 miles. Do cities A, B, C lie in a straight line?

   (1)   The distance between A and C is less than 35 miles.

   (2)   The distance between A and C is at least 15 miles.

B12:   Some quarters, dimes, and nickels add to $2.25. How many of each coin?

   (1)   There are 2 more quarters than dimes.

   (2)   There is one more nickel than quarters.

B13:   A taxi cost $4.00 for the first quarter of a mile and 50¢ for each additional quarter of a mile. How many miles is the trip?

   (1)   The trip cost $10.00.

   (2)   There is an additional $2.00 charge since the ride was after 10 p.m.

B14:   Is this a right triangle?

   (1)   The sides are in the ratio of $3 : 4 : 5$.

   (2)   Angles are in the ratio of $3 : 4 : 5$.

B15: Find a and b in the equation $x^2 + ax + b = 0$.

  (1) The sum of the roots is 6.

  (2) The product of the roots is 8.

B16: Is the area of the trapezoid greater than the area of the triangle?

  (1) The heights of both figures are equal, and the lower base of the trapezoid equals the base of the triangle.

  (2) Both perimeters are equal.

B17: Is the triangle an acute triangle (all angles less than 90°)?

  (1) The ratio of the angles is 5 : 6 : 7.

  (2) The triangle is isosceles.

B18: Given right triangle ABC with right angle C and hypotenuse AB. Find side AC.

  (1) $\angle B = 30°$.

  (2) $BC = 10$.

B19: Is the figure a rectangle?

  (1) The figure is a parallelogram.

  (2) The figure is a square.

B20: a, b, and c are positive integers. Is $ab + ca$ multiple of 7?

  (1) a is a multiple of 7.

  (2) c is a multiple of 7.

B21: x is a prime. Is $x + 1$ a prime?

  (1) $x + 2$ is a prime.

  (2) $x - 1$ is a prime.

B22: a, b are nonzero integers. Is $\dfrac{a}{b} > 1$?

  (1) $a > b$.

  (2) $a < 10$.

B23: Find the number of pounds of peanuts and the number of pounds of walnuts.

   (1)   Walnuts are $9 per pound and peanuts are $4 per pound.

   (2)   12 pounds of the mixture cost $86.

B24: Can we find y?

   (1) $x + 2y + 4z = 27$.

   (2) $2x + 8z = 50$.

B25: What is the value of x?

   (1) $-(x + y) = x - y$.

   (2) $x - y = 1$.

B26: Is $\dfrac{n - 48}{n}$ an integer?

   (1) $n^2 = 16$.

   (2) $n^2 = \dfrac{1}{4}$.

B27: If $x \neq 0$, the value of $\left(\dfrac{x^m}{x^n}\right)^3$

   (1) $m = n$.

   (2) $x = 10$.

B28: Does $x = 5$?

   (1) $x^2 = 25$.

   (2) $x^2 - 10x + 25 = 0$.

B29: x, y, z are positive integers. Does $y = \dfrac{x^2}{z}$?

   (1) $\dfrac{yz}{x^2} = 1$.

   (2) $x = \sqrt{yz}$.

B30: Is $a^b < b^a$, a, b are non-negative integers;

   (1) $a = b^2$.

   (2) $b > 3$.

B31: If $ab \neq 0$ does $\dfrac{1}{a} + \dfrac{1}{b} = 6$?

    (1) $6ab = a + b$.

    (2) $a = b$.

B32: If m is a positive integer, is $\sqrt{m}$ an integer?

    (1) $\sqrt{4m}$ is an integer.

    (2) $\sqrt{5m}$ is not an integer.

B33: x and y are integers and $y = |4 - x| + |5 + x|$; is $y = 9$?

    (1) $x > 1$.

    (2) $-1 < x < 5$.

B34: The cost of a phone call:

    (1)   It lasted 63 minutes.

    (2)   The cost for the first 3 minutes was 10 times the cost of each additional minute.

B35: The value of xy.

    (1) $y = x + 3$.

    (2) $y = x^2 + 3$.

B36: The arithmetic mean of m, n, and p.

    (1) $4m + 5n - 6p = 12$.

    (2) $2m + n + 12p = 60$.

B37: Determine if $x + \dfrac{1}{x} > y + \dfrac{1}{y}$.

    (1) $y < 0$ and $x > 0$.

    (2) $0 < y < x$.

Ⓐ **Let's look at the answers.**

B1:   **B:** Statement 1 is not sufficient; for example, if $a = 10$, $b = 8$, $c = 3$, and $d = 4$, then $a > d$; however, if $a = 10$, $b = 8$, $c = 3$, and $d = 4$, then $a < d$. Statement 2 is sufficient by the Law of Transitivity.

B2:   **C:** Individually, neither statement is sufficient because they each yield one equation with two unknowns: statement 1 gives $2^x 2^{3y} = 2^5$, or $x + 3y = 5$; statement 2 gives $3^x 3^{2y} = 3^4$, or $x + 2y = 4$. Together, however, if we subtract these equations, we get $y = 1$, and by substituting, we get $x = 2$.

B3:   **E:** Neither statement is sufficient. Even together they are not sufficient since we don't know whether there are more people involved or when each started, etc.

B4:   **A:** Statement 1 is sufficient: $e^3 = 6e^2$; so $e = 6$. Statement 2 is not sufficient; this fact is true for all cubes.

B5:   **A:** If $b = 0$, then $a\,b = (a)(0) + a + 0 = a$, irrespective of the value of a. This means that 0 is a right identity element. Thus statement 1 is sufficient. In fact, $a\,b = a$ implies that $a = ab + a + b$, which simplifies to $0 = ab + b = (b)(a + 1)$. The only allowable value of b is 0. Thus, statement 2 is not sufficient.

B6:   **E:** Statement 1 is not sufficient: In a plane, we have perpendicularity; if p comes out of the plane, it is not true. Statement 2 is not sufficient. In a plane, the lines are parallel. In the corner of any room, the statement is true. They cannot happen together.

B7:   **D:** In each case the smallest angle is $18°$.

B8:   **A:** Statement 2 is always true for all arithmetic progressions and is not sufficient. Statement 1 is true; using statement 2, $\dfrac{(a + 5a)}{2} = 7$; $6a = 14$; $a = \dfrac{7}{3}$; $5a = c = \dfrac{35}{3}$.

B9:   **B:** Statement 1 is not sufficient; $2^{n^2} = 2^4$; $n^2 = 4$; $n = \pm 2$. Statement 2 is sufficient; $2^{n^3} = 2^8$; $n^3 = 8$; $n = 2$.

B10:   **A:** Statement 1 is sufficient; increasing by 25% of the original is $\dfrac{5}{4}$ of the original; decrease by 20% means $\left(\dfrac{5}{4}\right)\left(\dfrac{4}{5}\right) = 1$ times the original; statement 2 is not sufficient; it is true only when $.9(n + 40) = n$.

B11: **E:** Neither statement is sufficient. Even together, we cannot determine if A, B, and C lie in a straight line. Here are some possible configurations:

I:

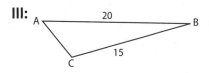

III:

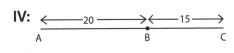

II:

IV:

B12: **C:** This solution would involve three equations in three unknowns. The original statement gives you one thing, the value. Statement 1 and statement 2 each give one more fact, but you need two more, so both statements together are sufficient. Remember, you don't have to solve the problem, but incidentally there are 6 quarters, 4 dimes, and 7 nickels.

B13: **A:** Statement 1 is sufficient. $\$10.00 = \$4.00 + (x - 1).50$, $x = \#$ of quarter miles. Statement 2 is insufficient.

B14: **D:** Statement 1 is sufficient; we have the classic 3-4-5 triangle, or a multiple of it. Statement 2 is sufficient because we have $3x + 4x + 5x = 180$ and $x = 15$; so 45, 60, and 75 shows it's not a right triangle.

B15: **C:** Individually, each statement is not enough. But together, as you may know, it is enough. Trial and error can show the roots are 4 and 2. The equation is $x^2 - 6x + 8 = 0$.

B16: **A:** Statement 1 is sufficient. The upper base makes the area of the trapezoid bigger. Statement 2 is not sufficient. You can make a trapezoid with a very small height, less area than an equilateral triangle or you can make a triangle with an itsy bitsy height and the trapezoid more like a square.

B17: **A:** Statement 1 is sufficient; $5x + 6x + 7x = 180$; the solution shows all the angles are less than 90°. Statement 2 is not sufficient. Angles could be 30-30-120 or 50-50-80.

B18: **C:** Neither is sufficient individually, but using both together, we get $AC = \dfrac{10}{\sqrt{3}}$.

B19: **B:** Statement 1 is not sufficient; not all parallelograms are rectangles. Statement 2 is sufficient; all squares are rectangles.

B20: **A:** $ab + ca = a(b + c)$. If a is a multiple of 7, then so is $a(b + c)$. Statement (2) is not sufficient. $a(b + c)$ need not be a multiple of 7 if c is a multiple of 7. For example, if $a = 3$, $b = 4$, $c = 7$, $a(b + c) = 33$, which is not a multiple of 7.

B21:   **D:**   Statement 1 is sufficient; if x and $x + 2$ are primes, they must be odd, and $x + 1$ is always even. Since $x + 1$ cannot be 2, $x + 1$ cannot be prime. Statement 2 is sufficient. If x and $x - 1$ are each prime, then x must be 3. This means that $x + 1 = 4$ cannot be prime.

B22:   **E:**   Neither is sufficient. Try $a = 5$, $b = 4$, and $a = -4$, $b = -5$; in each case, one works and the other doesn't.

B23:   **C:**   Statements 1 and 2 are not sufficient individually, but together they are sufficient; there are two equations with two unknowns: $9w + 4p = 86$ and $w + p = 12$.

B24:   **C:**   Individually neither is sufficient. However, combining the two, since $2x + 8z = 50$, $x + 4z = 25$; and using substitution, we get $x + 2y + 4z = 25 + 2y = 27$; so $y = 1$.

B25:   **A:**   Statement 1 is sufficient; $x = 0$; Statement 2 is insufficient.

B26:   **D:** 4, -4 and $\dfrac{1}{2}$, $-\dfrac{1}{2}$ make the expression an integer. Both statements are sufficient.

B27:   **A:**   Statement 1 is sufficient; its value is 1. Statement 2 is not sufficient; we need m and n.

B28:   **B:**   Statement 1 is not sufficient; we get $x = \pm 5$; Statement 2 is sufficient; we only get 5.

B29:   **D:**   Both statements are true individuals, no matter what the values of x, y, and z are.

B30:   **C:**   Statement 1 is not sufficient and is not true for $b = 1$, $b = 2$. Statement 2 is not sufficient; for example, $2^4 < 4^2$ is not true. However, both statements together make it true.

B31:   **A:**   Statement 1 is sufficient; we get $\dfrac{1}{a} + \dfrac{1}{b} = \dfrac{b + a}{ab} = 6$, which is equivalent to statement 1. Statement 2 is not sufficient. It is true if $a = \dfrac{1}{3}$, and not true if $a = 2$.

B32:   **A:**   Statement 1 is sufficient. If $\sqrt{4m} = 2\sqrt{m}$ is an integer, $\sqrt{m}$ must be an integer. Statement 2 is not sufficient; using $m = 4$, $\sqrt{20}$ is not an integer, but $\sqrt{4}$ is an integer. If $m = 7$, $\sqrt{35}$ is not an integer and neither is $\sqrt{7}$.

B33:   **B:**   Substitution shows that statement 1 is not sufficient, but statement 2 is sufficient. For example, if $x = 2$, the original equation is true; however, for $x = 6$, the original equation is false.

B34:   **E:**   Neither individually nor together are the statements sufficient since we don't know the specific rates.

B35:   **E:**   Neither is sufficient independently; when taken together, by substitution, we get two values for x, 0 and 1; which means two values for y, so together the statements are still insufficient.

B36: **C:** Neither statement is sufficient by itself; however, by adding the statements, we get $6m + 6n + 6p = 72$, And by dividing by 18, we get $\dfrac{m + n + p}{3} = 4$, the mean.

B37: **A:** Statement 1 is sufficient since a positive is always bigger than any negative. Statement 2 is not sufficient; $y = 4$, $x = 5$ is true; but $x = .2$ and $y = .1$ is false. Do not forget fractions if it doesn't say only integers!!

Ⓠ Data Sufficiency Practice Test C

A.   Statement (1) ALONE is sufficient, but statement (2) alone is not sufficient.

B.   Statement (2) ALONE is sufficient, but statement (1) alone is not sufficient.

C.   BOTH statements TOGETHER are sufficient, but NEITHER statement ALONE is sufficient.

D.   EACH statement ALONE is sufficient.

E.   Statements (1) and (2) TOGETHER are NOT sufficient.

C1: Determine the value of $\dfrac{a - b}{a + b}$.

   (1) $a = 4b$.

   (2) $a + b = 10$.

C2: $z = \sqrt{x - 7} + \sqrt{3 - y}$. What is the value of z?

   (1) $x = 11$.

   (2) $y = 2$.

C3: Is $x > y$?

   (1) $2 < y^3 < 65$;

   (2) $22 < x^2 < 65$.

C4: Find the angles of a parallelogram:

   (1) Opposite angles are equal.

   (2) Consecutive angles are supplementary.

C5: $[x]$ = largest integer $\leq x$; is $[x]$ larger than x?

   (1) $x = \pi$

   (2) $x = -3.5$

C6: a, b are integers ≥ 100. Is $\dfrac{a}{b} > 1$?

  (1) $|a - b| = 10$

  (2) $a + b = 300$

C7:   Is $\dfrac{2}{2x - 1} > 3$?

  (1) $1 \leq x \leq 2$.

  (2) $0 \leq x < .5$ or $.5 < x \leq 1$.

C8: A lives within 5 miles of B, and B lives within 5 miles of C. How close are A and C?

  (1) A lives within 5 miles of C.

  (2)   The distance between A and B is less than the distance between B and C.

C9: $y = \dfrac{a - b}{a + b}$; what is the value of y?

  (1) $a - b = 10$.

  (2) $a + b = 10$.

C10: $4a(5b) = x$; what is the value of x?

  (1) $b = a + 5$.

  (2) $2b = 32$.

C11:   What is 20% of y?

  (1)   60 is 4% of y.

  (2) $\dfrac{4}{5}$ of $y = 1200$.

C12:   The value of $x - y$.

  (1) $x = y + 10$.

  (2) $x^2 - 2xy + y^2 = 100$.

C13:   The value of x.

  (1) $2(x + 2) = -2x$.

  (2) $x^{11} = -1$.

C14: The number of rungs on a 20.5-foot ladder.

    (1) There is 9 inches between rungs and 9 inches between the ground and the first rung.

    (2) The width of the rung is 15.5 inches.

C15: n is an integer. Is $\dfrac{n}{5}$ an integer!

    (1) $\dfrac{3n}{5}$ is an integer.

    (2) $\dfrac{-4n}{5}$ is an integer.

C16: The value of x.

    (1) $x(x+2) = 8$.

    (2) $(2^x)^x = 16$.

C17: If $M = \dfrac{5a}{6b}$, $b \neq 0$; the value of M.

    (1) $b = 3a$.

    (2) $b = \dfrac{1}{3}$.

C18: $x^2 + kx + 24 = 0$; the value of k.

    (1) $(x - 4)$ is a factor of $x^2 + kx + 24 = 0$.

    (2) 6 is a root of this equation.

C19: The area of a 90° sector of a circle.

    (1) The sector's arc length is 10π.

    (2) $d = 40$.

C20: Is x a negative integer?

    (1) $3x > 5x$.

    (2) $x + 5$ a positive integer.

C21: a is what % of b?

    (1) $3a = 8b$.

    (2) $\dfrac{b}{a} = .375$.

C22: Find the value of x:

    (1) $x^2 = 9$.

    (2) $x = -|x|$.

C23: Is $x > y$?

    (1) $-3x < -3y$.

    (2) $x + z > y + z$.

C24: a, b, c are three angles of a triangle. What is the value of a?

    (1) $a + b = 70°$.

    (2) $b + c = 80°$.

C25: The amount of simple interest in one year.

    (1) The rate is 4%.

    (2) We have \$2,000 in the bank.

C26: The two-digit number tu, where t is the ten's digit and u is the unit's digit.

    (1) The sum of the digits is 6.

    (2) $t = u^2$.

C27: The length of a race on a circular track.

    (1) The race is 4 laps.

    (2) The area of the track is 1000 square meters.

C28: The ratio of $a : b : c$.

    (1) $a = 2$ and $bc = 60$.

    (2) $\dfrac{a}{b} = 3$ and $\dfrac{b}{c} = 4$.

C29: Is $2^n > 10$?

    (1) $2^{\sqrt{n}} = 4$.

    (2) $\dfrac{1}{2^n} < .1$.

C30: The value of prime p.

    (1) $31 < p < 41$.

    (2) $41 < p < 51$.

C31: Find the angles of a triangle.

    (1)   The measures of the angles are consecutive even integers.

    (2)   The difference between the first and second angles is $2°$ and the difference between the second and third angles is $2°$.

C32: Is m an integer?

    (1) $\dfrac{m}{3}$ is an integer.

    (2) $3m$ is an integer.

C33: $a < b$ and $c < d$.

    (1) $ac < bd$.

    (2) $a + c < b + d$.

C34: The total surface area of the sides of a rectangular box.

    (1)   The perimeter of the base is 20.

    (2)   The volume of the box is 2000.

C35: $x^M x^N = 64$; the value of x.

    (1) $N = 6 - M$.

    (2) $M = N = 3$.

C36: The length of a board cut in two pieces.

    (1)   The ratio of the two pieces is $5 : 2$.

    (2)   The difference in the pieces is 24 inches.

C37: $y = |2x - 4| + |2x + 4|$; the value of y.

    (1) $x^3 = 8$.

    (2) $x^2 = 4$.

A   **Let's look at the answers.**

C1:   **A:** Statement 1 is sufficient: $\dfrac{a-b}{a+b} = \dfrac{4b-b}{4b+b} = \dfrac{3}{5}$. Statement 2 is not sufficient:

$\dfrac{a-b}{a+b} = \dfrac{10-b-b}{10-b+b} = \dfrac{10-2b}{10} = \dfrac{5-b}{5}$ or $\dfrac{a-5}{5}$. We must know the value of

b (or a).

C2:   **C:** Separately, there is not enough information to get z, but if we substitute both statement 1 and statement 2, we get $z = 3$.

C3:   **E:** Neither statement is sufficient. If we look only at integers, statement 1 means $y = 2, 3,$ or 4; statement 2 says $x = 5, 6, 7, 8$ **and** $-5, -6, -7, -8$! So even together, the statements are insufficient.

C4:   **E:** Both statements are always true for all parallelograms; collectively, they do not help at all to find the specific angles of this parallelogram.

C5:   **D:** Both statements are sufficient, and the answer is no in each case. For statement 1, the bracket of pi is 3; for statement 2, the bracket of -3.5 is -4.

C6:   **E:** Neither equation alone is sufficient to determine which of a and b is the larger value. This means we cannot determine if $\dfrac{a}{b} > 1$ or $\dfrac{a}{b} < 1$. Taken together, the solutions are either $a = 155$ and $b = 145$ or $a = 145$ and $b = 155$. Thus we still cannot determine if $\dfrac{a}{b} > 1$ or if $\dfrac{a}{b} < 1$.

C7:   **A:** Statement 1 is sufficient; all values make the statement false. Statement 2 is not sufficient; $x = 0$ makes the statement false; but $x = 0.6$ makes the statement true.

C8:   **E:** Neither statement is sufficient, and collectively they don't help to establish distance between A and C.

C9:   **C:** Collectively the value of the fraction is 1; independently, neither statement is sufficient.

C10:   **C:** Statement 2 is not sufficient; it gives $b = 16$; statement 1 is not sufficient; however, together with statement 2, we get $a = 11$; and $20ab = (20)(11)(16) = 3520$. Again remember, you do not have to solve the problem; this solution is just FYI.

C11:   **D:** Statement 1 and 2 are both sufficient; each gives $y = 1500$; thus 20% of $y = 300$.

C12:   **A:** Statement 1 is sufficient: $x - y = 10$; Statement 2 is not sufficient: $(x - y)^2 = 100$ says $(x - y) = \pm 10$.

C13:   **D:** Statement 1 is sufficient: $4x + 4 = 0$; $x = -1$. Statement 2 is sufficient: again $x = -1$.

C14:   **A:** Statement 1 is sufficient. Statement 2 is insufficient.

C15:   **D:** If $\dfrac{3n}{5}$ is an integer, n must be a multiple of 5; so $\dfrac{3n}{5} = 3\left(\dfrac{n}{5}\right)$ must be an integer, similarly for $\dfrac{-4n}{5}$.

C16:   **C:** Statement 1 is not sufficient because we get the answers $x = 2$ and $x = -4$; statement 2 is not sufficient because we get $x = 2$ or $x = -2$. Taking the statement together, though, x must be 2.

C17:   **A:** Statement 1 is sufficient: $\dfrac{5a}{6b} = \dfrac{5a}{18a} = \dfrac{5}{18}$; statement 2 is not sufficient; $M = \dfrac{5a}{2}$, and a could be anything.

C18:   **D:** Statement 1 is sufficient to find $k = -10$; Statement 2 is also sufficient to find $k = -10$.

C19:   **D:** From statement 1 we get $c = 40\pi$, and thus $r = 20$. Statement 2 says $d = 40$, which is the same as $r = 20$. So both statements give $r = 20$, enough to find the area of the sector, $A = \dfrac{1}{4}\pi\,(20)^2$.

C20:   **A:** Statement 1 is sufficient. Statement 2 is not sufficient: for $x = -1$, the answer is yes, but for $x = 10$, it is no.

C21:   **D:** Both are sufficient and show a is $266\dfrac{2}{3}\%$ of b.

C22:   **C:** Statement 1 is not sufficient since $x = 3$ or -3. Statement 2 is not sufficient; it says $x = 0$ or negative. Together they are sufficient to determine $x = -3$.

C23:   **D:** Both are sufficient; they are both properties of inequalities.

C24:   **B:** Statement 1 is not sufficient; it gives the value for c. Statement 2 is sufficient: $a = 100°$.

C25:   **C:** Statement 1 is not sufficient because we need the amount. Statement 2 is not sufficient because we need the rate. Together, we have the amount and the rate, $I = prt = (2000)(.04)(1) = \$80$.

C26:   **C:** Statement 1 is not sufficient; the number could be 15, 24, 33, 42, or 51. Statement 2 is not sufficient; the number could be 11, 42, or 93. But together, we see that 42 is the answer.

C27: **C:** Statement 1 is not sufficient since we don't know the size of the track. Statement 2 is not sufficient; we can find r but not the number of laps. Together, though, they are sufficient.

C28: **B:** Statement 1 is not sufficient: $b = 2, c = 30$, or $b = 3, c = 20$ are two examples. Statement 2 is sufficient: $a = 3b, b = 4c$, so we have $a = 12c, b = 4c$, and $c = c$, which gives us $a : b : c = 12 : 4 : 1$.

C29: **D:** Both statements are sufficient: Statement 1 says $n = 4$. Statement 2 is equivalent because reciprocals of positives reverse the order of the inequalities.

C30: **A:** Statement 1 is sufficient: $p = 37$. Statement 2 is not sufficient: $p = 43$ or 47.

C31: **A:** Statement 1 is sufficient: $n + (n + 2) + (n + 4) = 180$. Statement 2 is not sufficient; it sounds the same, but it isn't; the first and third angles could be equal, or not equal (which would then be the same as statement 1).

C32: **A:** Statement (1) is sufficient, given that $\frac{m}{3}$ is an integer, $m = (3)\left(\frac{m}{3}\right) = (3)$ (an integer) must also be an integer. Statement (2) is not sufficient; if $3m = 7$, then $m = \frac{7}{3}$ which is not an integer.

C33: **B:** Let $a = 7, b = 8, c = 20, d = 22$. Then $a < b$ and $c < d$ would be true. But if $a = 20$, $b = 8, c = 7, d = 22$, then $a < b$ is false. This means statement (1) is not sufficient. For statement (2), again use $a = 7, b = 8, c = 20, d = 22$. Then $a < b$ and $c < d$ would be true. But by using $a = 20, b = 8, c = 7, d = 22$, then $a < b$ is false. This means statement 2 is not sufficient. Together, statements (1) and (2) are not sufficient.

C34: **E:** The dimensions of the base could be length $= 9$, width $= 1$, so that height $= 222\frac{1}{9}$. Total surface area of the sides $= (2)(1)\left(222\frac{1}{9}\right) + (2)(9)\left(222\frac{1}{9}\right) = 4444.\overline{4}$. But if length $= 8$, width $= 2$, and height $= 125$, the total surface area of the sides $= (2)(2)(125) + (2)(8)(125) = 2500$.

C35: **E:** Each statement gives $x^6 = 64$; $x = 2$ or $x = -2$. Neither statement is sufficient, alone or collectively.

C36: **C:** Statement 1 is not sufficient: the pieces are $5x$ and $2x$, but we don't know what x is. Statement 2 is not sufficient because it doesn't give the ratio of the pieces. Together, they are sufficient since $5x - 2x = 24$; $x = 8$; the board is $7x = 56$ inches.

C37: **D:** Very tricky: both are sufficient; whether $x = 2$ (statement 1) or $x = 2$ or -2 (statement 2), we get $y = 8$.

Now let's try some of the problems with which you are more familiar.

CHAPTER 16: *Problem Solving*

"*For total success, you must practice your skills.*"

Most of you have seen problems like these on the SAT. Let's do some.

Answer Sheets are at the back of the book.

Practice Test A

A1: If $x - 4 = 4 - x$, then $4x =$

    A.   0

    B.   4

    C.   8

    D.   16

    E.   256

A2: If $x = \sqrt[3]{-111}$, we know that x is

    A. $-11 < x < -10$

    B. $-10 < x < -9$

    C. $-5 < x < -4$

    D. $-4 < x < -3$

    E.   Undefined

A3: If $y = \dfrac{a^5}{b^3}$, then $\dfrac{ay}{b} =$

    A. $\dfrac{a^6}{b^4}$

    B. $\dfrac{a^4}{b^2}$

    C. $\dfrac{a}{b}$

    D. $\dfrac{a^{10}}{b^6}$

    E. $\dfrac{a^5}{b^3}$

A4: If $(x - 6)\left(6 - \dfrac{2}{x}\right) = 0$ and $x \neq 6$, then $x =$

    A. 0

    B. $\dfrac{1}{3}$

    C. $\dfrac{1}{2}$

    D. 2

    E. 3

A5: Jim does a job in 12 hours; Kim joins Jim and together they do the job in 4 hours. How many hours would it take Kim to do the job alone?

    A. 2

    B. 3

    B. 4

    D. 6

    E. 8

A6: *m* years ago *Cy* was *n* years old; how old was *Cy* *p* years ago?

A. $m + n - p$

B. $m - n - p$

C. $n - m - p$

D. $m - n - p$

E. $m + n + p$

A7: Which pairs are inverses of each other?

I: 3 and -3

II: 7 and $\dfrac{1}{7}$

III: $\sqrt{7}$ and $\dfrac{\sqrt{7}}{7}$

A. II only

B. II and III only

C. I and III only

D. I and II only

E. All of them

Questions A8–10 refer to the line $L: 3x - 4y = 7$

A8: The slope of L is

A. $\dfrac{7}{4}$

B. $\dfrac{3}{4}$

C. $\dfrac{4}{3}$

D. $-\dfrac{3}{4}$

E. $-\dfrac{4}{3}$

A9:   The *x*-intercept of *L* is:

A. $\left(\dfrac{7}{3}, 0\right)$

B. $\left(0, \dfrac{7}{3}\right)$

C. $\left(-\dfrac{7}{4}, 0\right)$

D. $\left(0, -\dfrac{7}{4}\right)$

E. $-\dfrac{7}{4}$

A10:   The *y*-intercept of *L* is:

A. $\left(\dfrac{7}{3}, 0\right)$

B. $\left(0, \dfrac{7}{3}\right)$

C. $\left(-\dfrac{7}{4}, 0\right)$

D. $\left(0, -\dfrac{7}{4}\right)$

E. $-\dfrac{7}{4}$

A11: $.9375 =$

A. $\dfrac{7}{8}$

B. $\dfrac{10}{11}$

C. $\dfrac{11}{12}$

D. $\dfrac{15}{16}$

E. $\dfrac{17}{18}$

A12: If x and y are integers, and at least one is odd, which of these expressions can never be odd?

    A. $x^2 + y^2$

    B. $2x^2 + y^2$

    C. $2xy + 1$

    D. $xy + 1$

    E. $x^2 + x + y^2 + y$

A13: $\dfrac{(6)(0.0048)}{0.72} =$

    A. 6

    B. 0.6

    C. .4

    D. 0.06

    E. 0.04

A14: $\dfrac{4 + 4\sqrt{7}}{4} =$

    A. $\sqrt{7}$

    B. $1 + 4\sqrt{7}$

    C. $1 + \sqrt{7}$

    D. $4 + \sqrt{7}$

    E. $4\sqrt{7}$

A15: The arithmetic mean of five numbers is 17; if one of the numbers is thrown away, the arithmetic mean is 18; the number thrown away is

    A. 16

    B. 15

    C. 14

    D. 13

    E. 12

A16: 7% of 48 is x% of 12; $x =$

    A. 14

    B. 21

    C. 28

    D. 42

    E. 56

A17: The area and the circumference of a circle have the same numerical value. The diameter of the circle is

    A. 1

    B. 2

    C. 3

    D. 4

    E. 5

A18: A fraction is equivalent to $\frac{2}{3}$; if 5 is added to the numerator, the fraction is equivalent to $\frac{5}{6}$. The original fraction was

    A. $\frac{54}{72}$

    B. $\frac{28}{42}$

    C. $\frac{20}{30}$

    D. $\frac{18}{27}$

    E. $\frac{16}{24}$

A19: What is the minimum number of sides you need to color a cube so that no two sides touching have the same color?

    A. 2

    B. 3

    C. 4

    D. 5

    E. 6

A20: If $\dfrac{y - x}{xy} = 2$, $y =$

    A. $\dfrac{x}{1 - x}$

    B. $\dfrac{x}{1 - 2x}$

    C. $\dfrac{x}{2 - x}$

    D. $\dfrac{2x}{1 - 2x}$

    E. $3x$

A21: Don has 30 coins in dimes and nickels that total $2.10. How many are dimes?

    A.   10

    B.   12

    C.   14

    D.   16

    E.   18

A22: $4^{2x} = 8^{x+5}$; $x =$

    A.   2

    B.   4

    C.   5

    D.   8

    E.   15

A23: $x^2 - y^2 = 28$ and $x + y = 2$; $x - y =$

    A.   4

    B.   7

    C.   14

    D.   21

    E.   Cannot be determined

A24: In an arithmetic progression, the first term, $a_1 = 15$, $a_2 = 9$, and $a_3 = 3$. Then $a_6 =$

   A. -3

   B. -9

   C. -15

   D. -21

   E. $\dfrac{1}{9}$

A25: Mel starts to read on the top of page 222 and ends reading at the bottom of page 358. The number of pages read is

   A.   136

   B.   137

   C.   138

   D.   139

   E.   580

A26: $x^2 - y^2 = 56$ and $x - y = 4$; $y =$

   A.   1

   B.   3

   C.   5

   D.   7

   E.   9

A27: $a \nabla b = ab^2 - b$. Then $-5 \nabla 3 =$

   A.   222

   B. -5

   C. -10

   D. -35

   E. -48

A28: $\sqrt{x-3} = 5; x =$

    A.   5

    B.   25

    C.   28

    D.   64

    E.   628

A29: $3x + 5$ is odd; the sum of the next two consecutive odd integers is

    A. $6x + 13$

    B. $6x + 14$

    C. $6x + 15$

    D. $6x + 16$

    E. $6x + 20$

A30: If x pounds of fruit cost c cents, how many pounds of fruit can you buy for d dollars?

    A. $\dfrac{100dx}{c}$

    B. $100cdx$

    C. $\dfrac{100c}{xd}$

    D. $\dfrac{cx}{100d}$

    E. $\dfrac{1}{100cdx}$

A31: $x + y = 12; x + \dfrac{x}{3} + y + \dfrac{y}{3} =$

    A.   4

    B.   8

    C.   12

    D.   16

    E.   20

A32: A jar has 5 yellow balls and 2 green balls; if two balls are picked one at a time, with replacement, the probability that both balls are green is

   A. $\dfrac{4}{7}$

   B. $\dfrac{4}{49}$

   C. $\dfrac{2}{21}$

   D. $\dfrac{1}{21}$

   E. $\dfrac{2}{49}$

A33: $3(2x - 1) + 4 = 2x; x =$

   A. $-\dfrac{1}{8}$

   B. $-\dfrac{1}{4}$

   C. $-\dfrac{1}{2}$

   D. $\dfrac{1}{4}$

   E. $\dfrac{1}{2}$

A34: In a small class, every girl sees that there is an equal number of girls and boys, while every boy sees there are twice as many girls as boys, not counting himself or herself. How many boys and how many girls are in the class?

   A.   3 boys and 3 girls

   B.   3 boys and 4 girls

   C.   4 boys and 3 girls

   D.   4 boys and 6 girls

   E.   2 boys and 3 girls

A35: Meg is 16 and her mother is 50. In how many years will Meg be half as old as her mother?

    A.   10

    B.   12

    C.   16

    D.   18

    E.   It can't happen

A36: If 3 bligs = 5 bloogs, and 7 bloogs = 8 blugs, the ratio of bligs to blugs is

    A. $\dfrac{21}{40}$

    B. $\dfrac{40}{21}$

    C. $\dfrac{105}{8}$

    D. $\dfrac{8}{105}$

    E. $\dfrac{21}{13}$

A37: Given the numbers 2, 2, 5, 7, 9, 17:   The sum of the median, mean and mode is

    A.   13

    B.   14

    C.   15

    D.   16

    E.   greater than 16

(A) **Let's look at the answers.**

A1:   **D:** $2x = 8$; $x = 4$; $4x = 16$.

A2:   **C:** $(-4)^3 = -64$; $(-5)^3 = -125$.

A3:   **A:** $y = \dfrac{a^5}{b^3}$, $\dfrac{ay}{b} = \dfrac{aa^5}{bb^3} = \dfrac{a^6}{b^4}$.

A4:   **B:** $6 - \dfrac{2}{x} = 0; \dfrac{2}{x} = 6; 6x = 2; x = \dfrac{1}{3}.$

A5:   **D:** $\dfrac{4}{12} + \dfrac{4}{x} = 1; \dfrac{4}{x} = \dfrac{2}{3}; 2x = 12; x = 6.$

A6:   **A:** The age now is the key; it is $m + n$; p years ago, it was $m + n - p$.

A7:   **E:** I is an additive inverse; II is a multiplicative inverse; III is a multiplicative inverse.

A8:   **B:** The slope-intercept form of the equation is $y = \left(\dfrac{3}{4}\right)x - \dfrac{7}{4}$. The slope is the coefficient of x.

A9:   **A:** $y = 0$ at the x-intercept, so $3x = 7$; $x = \dfrac{7}{3}\left(\dfrac{7}{3}, 0\right).$

A10: **D:** From the slope-intercept form of the equation, $y = (\dfrac{3}{4})x - \dfrac{7}{4}$, the y-intercept is $-\dfrac{7}{4}$.

A11: **D:** Write .9375 as $\dfrac{9375}{10,000}$. Divide numerator and denominator by 25 to get $\dfrac{375}{400}$, then reduce this fraction to $\dfrac{15}{16}$.

A12: **E:** If x is odd, x^2 is odd and the sum $(x + x^2)$ is even; if y is even, then y^2 is even, and the sum $(y + y^2)$ is even. The sum of evens is even. If you let x be even and y odd, all the others are odd. If both x, y are odd, then each of the four terms is odd and their sum is even.

A13: **E:** Divide first! 6 into .72 is .12 and .12 into .0048 = 0.04.

A14: **C:** $\dfrac{a + b}{c} = \dfrac{a}{c} + \dfrac{b}{c}$; so $\dfrac{4 + 4\sqrt{7}}{4} = \dfrac{4}{4} + \dfrac{4\sqrt{7}}{4} = 1 + \sqrt{7}.$

A15: **D:** Total is 5(17) = 85; 4(18) = 72; 85 − 72 = 13.

A16: **C:** Ignore the %; 7(48) = 12x; $x = \dfrac{7(48)}{12} = 7(4) = 28.$

A17: **D:** $\pi r^2 = 2\pi r$; so $r = 2$; and $d = 4$.

A18: **C:** $\dfrac{2x + 5}{3x} = \dfrac{5}{6}$; so 12x + 30 = 15x; 3x = 30, x = 10; thus $\dfrac{2x}{3x} = \dfrac{20}{30}.$

A19: **B:** Top and bottom can be one color; left and right can be another color; front and back can be another color.

A20: **B:** $y - x = 2xy$; $y - 2xy = x$; $y(1 - 2x) = x$; so $y = \dfrac{x}{1 - 2x}.$

A21:   **B:**   The easiest way to do this problem is to substitute each of the answer choices. If there are 10 dimes (answer choice A), then there would be 20 nickels, and the value would be 1.00 + 1.00 = 2.00. If there are 12 dimes (answer choice B), then there would be 18 nickels, and the value would be 1.20 + .90 = 2.10. That's the answer. Incidentally, the equation for solving this problem is $10x + 5(30 - x) = 210$, where x is the number of dimes.

A22:   **E:** $(2^2)^{2x} = (2^3)^{(x + 5)}$; so $2(2x) = 3(x + 5)$, or $4x = 3x + 15$; $x = 15$.

A23:   **C:** $x^2 - y^2 = (x - y)(x + y)$; $28 = (x - y)(2)$; $x - y = 14$.

A24:   **C:**   The difference is -6. So the terms of the progression are 15, 9, 3, -3, -9, -15.

A25:   **B:** $358 - 222 + 1 = 137$; note that you have to include the first page.

A26:   **C:**   Similar to problem A23, we now have $x + y = 14$; $x - y = 4$. By adding, we get $2x = 18$; $x = 9$; $y = 5$.

A27:   **E.** $(-5)(3^2) - 3 = -48$.

A28:   **C:**   Use trial and error with each of the answer choices. Alternatively, by squaring, $x - 3 = 25$, and $x = 28$.

A29:   **D:** $(3x + 7) + (3x + 9) = 6x + 16$. Note that even if the integers are odd, you must add 2 to get the next consecutive odd integer.

A30:   **A:** $\dfrac{x}{c} = \dfrac{?}{100d}$, or $? = \dfrac{100dx}{c}$.

A31:   **D:** $x + \dfrac{x}{3} + y + \dfrac{y}{3} = x + y + \dfrac{x}{3} + \dfrac{y}{3} = (x + y) + \dfrac{x + y}{3} = 12 + \dfrac{12}{3} = 16$.

A32:   **B:** $\left(\dfrac{2}{7}\right)\left(\dfrac{2}{7}\right) = \dfrac{4}{49}$; note that if there is no replacement, this changes to $\left(\dfrac{2}{7}\right)\left(\dfrac{1}{6}\right) = \dfrac{1}{21}$.

A33:   **B:** $6x - 3 + 4 = 2x$; $4x = -1$; $x = -\dfrac{1}{4}$.

A34:   **B:**   Each girl sees 3 girls and 3 boys; each boy sees 4 girls and 2 boys.

A35:   **D:** $\dfrac{(x + 16)}{(x + 50)} = \dfrac{1}{2}$; so $2x + 32 = x + 50$; $x = 18$.

A36:   **B:** $\dfrac{\text{bligs}}{\text{bloogs}} = \dfrac{5}{3}$ and $\dfrac{\text{bloogs}}{\text{blugs}} = \dfrac{8}{7}$; therefore, $\dfrac{\text{bligs}}{\text{blugs}} = \dfrac{5}{3} \times \dfrac{8}{7} = \dfrac{40}{21}$.

A37:   **C:** Median is 6; Mode is 2; Mean is $\dfrac{42}{6} = 7$.

Practice Test B

B1: m and n are odd integers: Which are always odd?

   I: m^{n+1}

   II: $mn + m + n$

   III: $(m - 2)^{n-4}$

   A.   None

   B.   I and II only

   C.   I and III only

   D.   II and III only

   E.   I, II, and III

B2:   If y is four less than the square root of x, and $x > 100$, then

   A. $x = 4 + y^2$

   B. $x = 4 - y^2$

   C. $x = y^2 - 4$

   D. $x = (y - 4)^2$

   E. $x = (y + 4)^2$

B3:   .3% of .3% of .3 =

   A.   .027

   B.   .0027

   C.   .00027

   D.   .000027

   E.   . 0000027

B4: If $-1 < x < 0$, which of the following is arranged in order, largest to smallest?

A. $x^5 > x^4 > x^3$

B. $x^3 > x^4 > x^5$

C. $x^5 > x^3 > x^4$

D. $x^4 > x^3 > x^5$

E. $x^4 > x^5 > x^3$

B5: At 60° F, you need 15 SPF sunscreen. For each 5-degree increase, the SPF must increase by 3. At 95° F, the required SPF is

A. 30

B. 35

C. 36

D. 40

E. 50

B6: In a large lecture hall, the ratio of men to women is 2 : 3. If there are 365 people in the lecture hall, the number of women is

A. 73

B. 105

C. 126

D. 155

E. 219

B7: A circle has area 1. Its diameter is

A. $\dfrac{1}{\pi}$

B. $\dfrac{2}{\pi}$

C. $\dfrac{1}{\sqrt{\pi}}$

D. $\dfrac{2}{\sqrt{\pi}}$

E. π

B8:   The volume of a cube whose surface area is 6 is

    A.   1

    B.   6

    C.   36

    D.   216

    E.   1996

B9: $\dfrac{x}{2} + \dfrac{x}{3} = 1; x =$

    A. $\dfrac{2}{3}$

    B. $\dfrac{3}{2}$

    C. $\dfrac{6}{5}$

    D. $\dfrac{9}{4}$

    E. $\dfrac{11}{10}$

B10: $2^n + 2^n + 2^n + 2^n =$

    A. 2^{n+1}

    B. 2^{n+2}

    C. 2^{n+4}

    D. 4^n

    E. 2^{n^4}

B11: $\dfrac{\left(5x^4\right)^2 x^3}{5x^4} =$

    A. x^3

    B. $5x^3$

    C. x^7

    D. $5x^7$

    E. x^{15}

B12: $x^6 = m; x^5 = \dfrac{n}{3};$ in terms of m and n, $x =$

    A. $3mn$

    B. $\dfrac{3}{mn}$

    C. $\dfrac{mn}{3}$

    D. $\dfrac{3m}{n}$

    E. $\dfrac{3n}{m}$

B13: A $60 radio is discounted 25%, then another 10%; with a 4% sales tax, the final cost is

    A. $41.06

    B. $42.12

    C. $43.18

    D. $44.24

    E. $45.30

B14: a is a positive integer; a^2 ends in a 9; $(a + 1)^2$ ends in a 4; $(a + 2)^2$ ends in

    A.   1

    B.   4

    C.   5

    D.   6

    E.   9

B15: Set $R = \{1, 3, 5, 7\}$ and $R \cup S = \{1, 2, 3, 4, 5, 6, 7\}$. S can be which one of the following sets?

    A. $\{1, 4, 6\}$

    B. $\{2, 4, 6\}$

    C. $\{1, 2, 3, 4, 5, 6, 7, 8\}$

    D. $\{1, 3, 5, 7\}$

    E. $\{1, 2, 3, 4\}$

Any two sides

B16:   Two sides of a right triangle are $13\sqrt{2}$ and 13. The third side could be:

   A.   only 13

   B.   only $13\sqrt{2}$

   C.   only $13\sqrt{3}$

   D.   either 13 or $13\sqrt{3}$

   E.   need to know the angles to answer the problem

B17: $\sqrt{.000036}$ is exactly

   A.   .00006

   B.   .0006

   C.   .006

   D.   .06

   E.   Not an exact number

Problems B18 and B19 refer to a Norman window, which is a rectangle surmounted by a semicircle. The height of the rectangular part is x and the width of the rectangular part is $4y$.

B18:   The perimeter is

   A. $2x + 8y + 4\pi y$

   B. $2x + 4y + 4\pi y$

   C. $2x + 4y + 2\pi y$

   D. $2x + 4y + \pi y$

   E. $2x + y + \dfrac{\pi y}{2}$

B19:   The area is

   A. $xy + \dfrac{\pi y^2}{2}$

   B. $4xy + 4\pi y^2$

   C. $4xy + 2\pi y^2$

   D. $4xy + \pi y^2$

   E. $4xy + \dfrac{\pi y^2}{4}$

B20: If $\dfrac{x-6}{y+4} = 0$, then

    A. $x = 0, y \neq 0$

    B. $x = 6, y = -4$

    C. $x = 6, y \neq -4$

    D. $x \neq 6, y \neq -4$

    E. $x \neq 6, y \neq -4$

B21: If $\dfrac{2}{5}$ of n is $\dfrac{7}{3}$; $\dfrac{2}{7}$ of n is

    A. $\dfrac{3}{5}$

    B. $\dfrac{5}{3}$

    C.   6

    D.   35

    E.   210

B22: An item cost \$600 after a 25% discount. Originally the cost was

    A.   \$150

    B.   \$200

    C.   \$700

    D.   \$750

    E.   \$800

B23: A and B are on opposite sides of point P on a line and $3AP = 4PB$. M is the midpoint of AP. $AM : MB =$

    A. $1 : 4$

    B. $2 : 7$

    C. $1 : 3$

    D. $2 : 5$

    E. $4 : 9$

B24: Simplified, $\dfrac{\frac{1}{m^2} - \frac{1}{n^2}}{\frac{1}{m} - \frac{1}{n}} =$

A.   1

B. $\dfrac{1}{mn}$

C. $\dfrac{m+n}{mn}$

D. $mn + 1$

E. $\dfrac{1}{mn + 1}$

B25: $100^M = 10^{100}$; $M =$

A.   10

B.   20

C.   25

D.   50

E.   75

B26: $(x - 7)^2 = (x - 19)^2$; $x =$

A.   9

B.   11

C.   13

D.   15

E.   17

B27: If $c + d - 6 = 8g$, which is the mean of c, d, and g?

A. $3g - 2$

B. $3g$

C. $3g + 2$

D. $2g + 1$

E. $2g - 1$

B28: $abc = bcde > 0$, a, b, c, d, e are all integers > 1. Which of the following *must* be true?

    A. $a > d$

    B. $b > c$

    C. $c > e$

    D. $d > b$

    E. $e > a$

B29: Given the sequence $-1, 0, 1, -1, 0, 1, \ldots$ the sum of the 82nd term and 112th term is

    A. -2

    B. -1

    C. 0

    D. 1,

    E. 2

B30: $a \# b = b + a^3$; $2 \# x^2 = 9x$ has solutions

    A. $x = 0$ and 1 only

    B. $x = 0$ and 8 only

    C. $x = 1$ and 8 only

    D. $x = 0, 1$, and 8 only

    E. It involves a 6th-degree equation, which is unsolvable

B31: The sum of seven consecutive odd integers is -105. The product of the two largest is

    A. 99

    B. 182

    C. 225

    D. 206

    E. 399

B32: Max goes 80 km/hr in one direction and 100 km/hr on the return trip on the same road. The average speed to the nearest tenth of a km is

A.   85.3

B.   86.7

C.   88.9

D.   90.0

E.   92.3

B33: $M = 2^r$; $8M =$

A. 16^r

B. 64^r

C. 2^{r^3}

D. 2^{r+3}

E. 2^{3r}

B34: The maximum value of $\dfrac{m + n}{m - n}$ with $8 \leq m \leq 10$ and $2 \leq n \leq 4$ is

A. $\dfrac{3}{2}$

B. $\dfrac{5}{3}$

C. $\dfrac{7}{3}$

D.   3

E.   6

B35: $0 < m < 1$

I: $m < \dfrac{1}{\sqrt{m}}$

II: $\dfrac{1}{m} < \dfrac{1}{m^2}$

III: $1 - m \leq m$.

Which statement(s) is(are) always true?

A.   None

B.   I and II only

C.   I and III only

D.   II and III only

E.   All are always true

B36: $\dfrac{x + 1}{x - 1} = \dfrac{x - 1}{x + 1}$; The solution ($x$) is (are)

A. $x = 0$ only

B. $x = 0, 1$ only

C. $x = 0, -1$ only

D. $x = 0, 1, -1$ only

E.   There are no solutions

B37: $\dfrac{3\frac{2}{3} - 1\frac{2}{3}}{\frac{5}{3} - \frac{1}{3}} =$

A.   1

B. $\dfrac{3}{2}$

C.   2

D. $\dfrac{5}{2}$

E. $\dfrac{8}{3}$

Ⓐ **Let's look at the answers.**

B1: **B:** I is always odd; an odd integer to an even power is odd. II is odd since the sum of 3 odd integers is always odd. III is not always odd; for example, let $m = 5$ and $n = 1$.

B2: **E:** The equation is $y = \sqrt{x} - 4$, or $y + 4 = \sqrt{x}$, which is $(y + 4)^2 = x$. The fact that $x > 100$ is superfluous to the solution.

B3: **E:** $(.003)(.003)(.3) = .0000027$.

B4: **E:** x^4 is largest since it is the only positive. $\left(-\dfrac{1}{2}\right)^5 > \left(-\dfrac{1}{2}\right)^3$ since it is closer to 0.

B5: **C:** $95 - 60 = 35$, Thus, there are 7 groups of 5; $7(3) = 21$ increase in SPF; $21 + 15 = 36$.

B6: **E:** $2x + 3x = 365; x = \dfrac{365}{5} = 73; 3x = 219$.

B7: **D:** Area $= 1 = \pi r^2; r = \dfrac{1}{\sqrt{\pi}}; d = 2r = \dfrac{2}{\sqrt{\pi}}$.

B8: **A:** $6e^2 = 6; e^2 = 1; e^3 = 1$.

B9: **C:** Multiply the equation by 6 to get $5x = 6$. Then $x = \dfrac{6}{5}$.

B10: **B:** This is a tough one! $2(2^n + 2^n) = 2(2)(2^n) = (2^2)(2^n) = 2^{n+2}$.

B11: **D:** $\dfrac{25x^8(x^3)}{5x^4} = 5x^7$.

B12: **D:** $\dfrac{x^6}{x^5} = \dfrac{m}{\left(\frac{n}{3}\right)} = \dfrac{3m}{n}$.

B13: **B:** \$60 discounted 25% = \$45 discounted 10% $(.90 \times 45) = \$40.50 \times .04 = \$1.62 + \$40.50 = \$42.12$.

B14: **A:** $0^2 = 0; 1^2 = 1; 2^2 = 4; 3^2 = 9; 4^2 = 16; 5^2 = 25; 6^2 = 36; 7^2 = 49; 8^2 = 64; 9^2 = 81$. Since a^2 ends in a 9, a must end in a 3 or 7, but $(a + 1)^2$ ends in a 4, so a must end in a 7. Thus, $a + 2$ ends in a 9, and finally $(a + 2)^2$ must end in a 1.

B15: **B:** Each element in the union of two sets is an element in one or the other or both sets, and all elements in either set must be included in the union.

B16: **D:** Use the Pythagorean Theorem. The missing side could be the hypotenuse $(13\sqrt{3})$ or one of the legs (13).

B17: **C:** The square root has half the decimal places as the radicand (quantity under the root sign).

B18: **C:** $p = \left(\frac{1}{2} \times 2\pi r\right) + x + x + 4y; r = 2y; p = 2\pi y + 2x + 4y.$

B19: **C:** Rectangle is $bh = 4xy$; semicircle $\left(\frac{1}{2}\right)\pi(2y)^2 = 2\pi y^2$. $A = \frac{1}{2}\pi r^2 + bh = \frac{1}{2}\pi(2y)^2 + 4y(x) = 4xy + 2\pi y^2$.

B20: **C:** A fraction equals 0 if the top equals 0 and the bottom does not equal 0.

B21: **B:** $\frac{2}{5}n = \frac{7}{3}$; so $\frac{2}{7}n = \frac{5}{3}$. Exchange the 5 and 7; just another pattern of fractions.

B22: **E:** Plug numbers into the answer choices, or use $x - .25x = 600$, so $\frac{3}{4}x = 600$, and $x = 800$.

B23: **D:** Hard to picture in your mind. Let $AB = AP + PB$, where $AP = 4$ (equal intervals) and $PB = 3$. Thus, AP is divided into 7 equal parts. Since M is the midpoint of AP, $AM = 2$, and $MB = 7 - 2 = 5$; so the ratio $AM : MB = 2:5$.

B24: **C:** Multiplying by m^2n^2 and factoring, we get $\frac{(m+n)(m-n)}{mn(m-n)} = \frac{m+n}{mn}$.

B25: **D:** $10^{2M} = 10^{100}$; $2M = 100$, so M $= 50$.

B26: **C:** Trial and error with each answer choice gives $(13 - 7)^2 = (13 - 19)^2$, or $(6)^2 = (-6)^2$. Note that 13 is the mean of 7 and 19.

B27: **C:** $\frac{c+d+g}{3}$. If $c + d - 6 = 8g$, then $c + d = 8g + 6$. Substituting, we get the

mean $= \frac{8g + 6 + g}{3} = 3g + 2.$

B28: **A:** b and c are unknown $a = de$; a, b, e are positive integers; so a is larger than either d or e.

B29: **A:** 82nd and 112th numbers are both the first number of the three-number repeating sequence $-1, 0, 1$, so divide each number by 3, they each have a remainder of 1, so they both end in -1, and therefore their sum is -2.

B30: **C:** $x^2 + 8 = 9x$ or $x^2 - 9x + 8 = 0$ or $(x - 8)(x - 1) = 0$.

B31: **A:** The middle term is the average, or $\frac{-105}{7} = -15$; The integers are $-21, -19, -17,$ $-15, -13, -11, -9$, and the product of the two largest is $(-11)(-9) = 99$.

B32: **C:** Suppose the trip was 400 km each way. Then the times of travel $\left(t = \frac{d}{r}\right)$ are $\frac{400}{80} = 5$ hours in one direction and $\frac{400}{100} = 4$ hours in the other direction. The rate for the entire trip is total distance divided by total time, or $\frac{800}{9} = 88.9$.

B33: **D:** $8M = 8(2^r) = 2^3 2^r = 2^{r+3}$.

B34: **D:** In this case $m = 8$ and $n = 4$ works. Substituting only the end numbers, $\dfrac{8 + 4}{8 - 4} = 3$.

B35: **B:** Statement I is true since $m < 1$ and $\dfrac{1}{\sqrt{m}} > 1$; Statement II is true since $m^2 < m$; so

$\dfrac{1}{m^2} > \dfrac{1}{m}$; Statement III is false (try $m = \dfrac{3}{4}$, for example.)

B36: **A:** By cross multiplying, we get $4x = 0$, so $x = 0$ only.

B37: **B:** $\dfrac{\frac{6}{3}}{\frac{4}{3}} = \dfrac{6}{4} = \dfrac{3}{2}$

Practice Test C

C1: The next term in the sequence 0, 1, 1, 2, 3, 5, 8, 13,…is

    A. 18

    B. 19

    C. 21

    D. 23

    E. 25

C2: $(-3)^{2m} = 3^{9-m}$, m an integer; $m =$

    A. 1

    B. 2

    C. 3

    D. 4

    E. 5

C3: $x^2 + kx - 24 = (x + 3)(x + m)$; $k =$

    A. 8

    B. 5

    C. -5

    D. -8

    E. -13

C4: $\dfrac{m}{n} = \dfrac{3}{4}$. Which is NOT true?

A. $\dfrac{3m}{4n} = \dfrac{9}{16}$

B. $\dfrac{(m + n)}{n} = \dfrac{7}{4}$

C. $\dfrac{(n - m)}{m} = \dfrac{1}{3}$

D. $\dfrac{n^2}{m^2} = \dfrac{16}{9}$

E. $\dfrac{(m - n)}{n} = \dfrac{1}{4}$

C5:   All 35 students take at least French or Russian. If 25 take Russian and 10 take both, how many students take only French?

A.   5

B.   10

C.   15

D.   20

E.   25

Problems C6 and C7 refer to the following figure, in which AB ∥ CD.

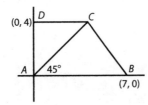

C6:   The area of ABCD =

A.   11

B.   16

C.   22

D.   32

E.   56

C7:   The perimeter of *ABCD* =

    A.   20

    B.   22

    C.   25

    E.   40

    E.   140

C8: $0 < x < 1$: Which must be true?

    I: $x^9 < x^7$

    II: $x^7 + x^5 < x^6 + x^4$

    III: $x^7 - x^5 < x^6 - x^4$

    A.   I only

    B.   II only

    C.   III only

    D.   I and II only

    E.   I, II, and II

C9:   In the figure below, twice the area of $\triangle XYZ$ =

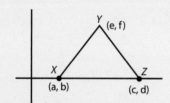

    A. $(d - b)e$

    B. $(c - a)f$

    C. $(c - a)e$

    D. $\frac{1}{2}(c - a)f$

    E. $\frac{1}{2}(d - b)e$

C10: The closest approximation to $\sqrt{\dfrac{(8.97)(902.1)}{35.97}}$ is

   A. 5

   B. 15

   C. 30

   D. 225

   E. 900

C11: A football team averaged x points (the mean) in the first n games, scored y points in the next game, and z points in the one after that. The mean of the $n + 2$ games is

   A. $\dfrac{x + y + z}{n + 2}$

   B. $\dfrac{nxyz}{n + 2}$

   C. $\dfrac{nx + y + z}{n + 2}$

   D. $x + \dfrac{y + z}{2}$

   E. $\dfrac{nx + yz}{n + 2}$

C12: The postage in country X is 50 gruff for the first ounce and 22 gruff for each additional ounce. A half-pound letter mailed in country X cost how many gruff?

   A. 204

   B. 226

   C. 248

   D. 270

   E. 292

C13: Three business partners, U, V, and W, divided the profits, respectively, in a ratio of $5 : 7 : 8$. If W's share was $200,000, the total profit of the business was

   A. $100,000

   B. $200,000

   C. $250,000

   D. $500,000

   E. $1,000,000

C14: In the figure below, the largest side is

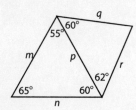

A. m

B. n

C. p

D. q

E. r

C15: Two sides of a triangle are 4 and 7. If the sides are integers, the product of the smallest possible perimeter and largest possible perimeter is

A. 17

B. 36

C. 40

D. 210

E. 315

C16: j, k, and m are three consecutive odd integers, with $j < k < m$; which is true?

I: $m - j = 4$

II: $jk + jm + km + 1$ is even

III: $\dfrac{(j + k + m)}{3}$ is an integer

A. I only

B. I and II only

C. I and III only

D. II and III only

E. I, II, and III

C17: The sum of the three smallest primes greater than 30 is

    A. 99

    B. 109

    C. 111

    D. 115

    E. 131

C18: If the following were written as decimals, which would have the longest repeating sequence of different digits?

    A. $\dfrac{1}{3}$

    B. $\dfrac{1}{6}$

    C. $\dfrac{1}{7}$

    D. $\dfrac{1}{9}$

    E. $\dfrac{1}{99}$

C19: The perimeter of a sector of $40°$ in a circle whose circumference is 36π is

    A. $2\pi + 18$

    B. $2\pi + 36$

    C. $4\pi + 18$

    D. $4\pi + 36$

    E. $4\pi + 72$

C20: Which is not equivalent to $36a^2 = b^2 - 25$?

    A. $(b + 6a)(b - 6a) = 25$

    B. $b^2 = (6a + 5)(6a - 5)$

    C. $\dfrac{(b^2 - 25)}{36a^2} = 1$

    D. $-a^2 = \dfrac{(5 - b)(5 + b)}{36}$

    E. $36a^2 + 25 - b^2 = 0$

C21:   Which of the following equations has a root in common with $x^2 - 7x + 6 = 0$?

    A. $x^2 + 36 = 0$

    B. $x^2 + 1 = 0$

    C. $x^2 + 7x + 6 = 0$

    D. $x^2 - 5x - 6 = 0$

    E. $x^2 + 8x + 12 = 0$

C22:   The largest factor of 3,000,000,036 listed here is

    A.   2

    B.   6

    C.   12

    D.   132

    E.   396

C23: $\sqrt{6561}$ is exactly

    A.   81

    B.   83

    C.   89

    D.   91

    E.   99

C24: $\dfrac{x}{4} + \dfrac{x}{8}$ is what percent of x?

    A. $33\dfrac{1}{3}$

    B. $37\dfrac{1}{2}$

    C.   40

    D. $43\dfrac{3}{4}$

    E.   45

C25: If $x > 10,000$, $\dfrac{2x}{3x + 5}$ is approximately

A. $\dfrac{2}{3}$

B. $\dfrac{2}{5}$

C. $\dfrac{1}{4}$

D. $\dfrac{1}{5}$

E. $\dfrac{1}{15}$

C26: According to the chart below, the median grade was

Number of Students	Grade
10	60
64	70
35	80
25	90
15	100

A. 70

B. 75

C. 80

D. 85

E. Can't be determined

C27: $\dfrac{6}{x} = 2$ and $\dfrac{12}{y} = 24$; $\dfrac{3x + 1}{y + 2} =$

A. $\dfrac{5}{2}$

B. 3

C. $\dfrac{7}{2}$

D. 4

E. 6

C28:   Which fraction has the greatest value?

A. $\dfrac{6}{2^2 5^2}$

B. $\dfrac{8}{2^3 5^2}$

C. $\dfrac{12}{2^3 5^3}$

D. $\dfrac{18}{2^4 5^3}$

E. $\dfrac{112}{2^4 5^4}$

C29:   If c and d are different integers and $c^2 = cd$, which must be true?

I. $c = -d$

II. $c = 0$

III. $d = 0$

A.   I only

B.   II only

C.   III only

D.   II and III only

E.   I, II, and III

C30:   How many square inches of wrapping paper will Kara need to just cover a gift box that measures 3 inches by 5 inches by one foot?

A.   15 square inches

B.   36 square inches

C.   60 square inches

D.   111 square inches

E.   222 square inches

C31: How many nonnegative integers x are there such that $x^4 < 1{,}000$?

A. 2

B. 3

C. 4

D. 5

E. More than 5

C32: In the figure below, which line is not part of the boundary?

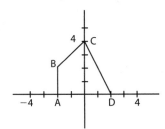

A. $x = -2$

B. $y = 0$

C. $y - x = 4$

D. $2x + y = 4$

E. $x + 2y = 4$

C33: $(\sqrt{2} - 1)(\sqrt{2} + 1)(\sqrt{3} + 1)(\sqrt{3} - 1)(\sqrt{5} - 1)(\sqrt{5} + 1) =$

A. $\sqrt{30}$

B. 7

C. 8

D. 16

E. 30

C34: A certain virus triples every 15 minutes. Originally there are 10,000 viruses. How many viruses are there in one hour?

A. 30^4

B. 10^8

C. 30^8

D. $3^4 10^8$

E. 30^{16}

C35: The perimeter of a small garden is 24 feet and its area is 20 square feet. The length of the shorter side is

A. 2 ft

B. 4 ft

C. 6 ft

D. 8 ft

E. 10 ft

C36: How many total ounces are there in $\frac{1}{2}$ gallon, 1 quart, plus $1\frac{1}{2}$ pints?

A. 100 ounces

B. 110 ounces

C. 120 ounces

D. 128 ounces

E. 132 ounces

C37: To convert from Fahrenheit (F) temperature to Celsius (C), the formula is $F = \frac{9}{5}C + 32$. Where do the two scales have the same temperature?

A. $20°$

B. $-20°$

C. $-40°$

D. $-100°$

E. $-273°$

(A) **Let's look at the answers.**

C1: **D:** Add the last two terms to get the next one: $0 + 1 = 1; 1 + 1 = 2; 1 + 2 = 3; 2 + 3 = 5;$ $3 + 5 = 8; 5 + 8 = 13; 8 + 13 = 21.$

C2: **C:** The minus sign doesn't matter since both sides are positive; $2m = 9 - m; m = 3.$

C3: **C:** $3m = -24; m = -8; 3x + mx = 3x + -8x = -5x;$ so $k = -5.$

C4: **E:** $\dfrac{(m - n)}{n} = \dfrac{m}{n} - \dfrac{n}{n} = \dfrac{m}{n} - 1 = \dfrac{3}{4} - 1 = -\dfrac{1}{4}.$

C5: **B:** $N(F) + N(R) - N(both) = N(F \text{ or } R)$; $N(F) + 25 - 10 = 35$; $N(F) = 20$; but $N(F) = 20$ and $N(both) = 10$; so $N(F \text{ only}) = 20 - 10 = 10$.

C6: **C:** Point $C = (4, 4)$; Area $= \left(\dfrac{1}{2}\right)h(b_1 + b_2) = \left(\dfrac{1}{2}\right)AD(AB + CD) = \left(\dfrac{1}{2}\right)(4)(7 + 4) = 22$.

C7: **A:** $p = AD + CD + AB + BC = 4 + 4 + 7 + \sqrt{(4-7)^2 + (4-0)^2} = 15 + 5 = 20$.

C8: **D:** I is true; II is true because $x^5(1 + x^2) < x^4(1 + x^2)$ since $x^5 < x^4$; III is false. $x^7 - x^5 = x^5(x^2 - 1)$ and $x^6 - x^4 = x^4(x^2 - 1)$. We know that $x^5 < x^4$ when $0 < x < 1$. However, $x^2 - 1$ is a negative number for $0 < x < 1$. Thus, $x^5(x^2 - 1)$ is actually greater than $x^4(x^2 - 1)$. Using $x = .5$, $x^7 - x^5 \approx -.023$ and $x^6 - x^4 \approx -.047$.

C9: **B:** Remember, it asks for twice the area; only on a test like this one! You must read carefully. Notice that $b = d = 0$, since points x and z lie on the x axis.

C10: **B:** The given square root is approximately $\sqrt{\dfrac{9(900)}{36}} = \dfrac{3(30)}{6} = \dfrac{30}{2} = 15$.

C11: **C:** In the first n games, an average of x means nx total points. Then add the next two games. Mean $= \dfrac{\text{total points}}{\text{total games}} = \dfrac{nx + y + z}{n + 2}$.

C12: **A:** 16 ounces to a pound, and 8 ounces to a half pound. The first ounce cost 50 gruff, and the remaining 7 cost 22 gruff each: $50 + 7(22) = 204$ gruff.

C13: **D:** $5x + 7x + 8x = 20x$, the total; $8x = \$200{,}000$; $4x = \$100{,}000$; $20x = 5(4x) = \$500{,}000$.

C14: **D:** In the left triangle, p is the largest side; in the right triangle, q is the largest side; but p and q are in the same triangle, so q is the largest side.

C15: **E:** The missing side s: $7 - 4 < s < 7 + 4$. Since they are all integers s could be 4 through 10. The smallest perimeter is $4 + 4 + 7 = 15$. The largest is $10 + 4 + 7 = 21$. The product is $15 \times 21 = 315$. In this case, use approximation to see that the answer must be larger than $15 \times 20 = 300$, and the only answer choice is 315.

C16: **E:** The three consecutive odd numbers can be represented as j, $j + 2$, and $j + 4$. Statement I is true, since $j + 4 - j = 4$. Statement II is true, since (odd)(odd) + (odd)(odd) + (odd)(odd) + 1 = odd + odd + odd + 1 = even. Statement III is true, since $\dfrac{j + (j + 2) + (j + 4)}{3} = \dfrac{3j + 6}{3} = j + 2$, which is the middle integer K.

C17: **B:** $31 + 37 + 41 = 109$.

C18: **C:** $\frac{1}{7}$ repeats every 6 places; all the others repeat a single digit, except $\frac{1}{99}$ which repeats 2 digits.

C19: **D:** $p = 2r + s; c = 36\pi = 2\pi r; r = 18; s = \frac{40}{360}(2\pi r) = \frac{1}{9}(36\pi) = 4\pi; p = 36 + 4\pi.$

C20: **B:** $b^2 = 36a^2 + 25$, which does not factor.

C21: **D:** Both have root $x = 6. x^2 - 7x + 6 = 0$ factors as $(x - 6)(x - 1) = 0$ and $x^2 - 5x - 6 = 0$ factors as $(x - 6)(x + 1) = 0.$

C22: **D:** The sum of the digits is 12, which is divisible by 3 (but not 9); The sum of the odd digits = sum of the even digits (6 = 6); so the number is divisible by 11. The number ends in 36, which is divisible by 4; 4(11)(3) = 132. 396 is not okay since 9 is one of its factors.

C23: **A:** $\sqrt{6400} = 80$; the number ends in 1; so the answer must be 81 or 89. The number is closer to 6400 than 8100.

C24: **B:** $\frac{x}{4} + \frac{x}{8} = \frac{3}{8}x$; and $\frac{3}{8} = 37\frac{1}{2}\%.$

C25: **A:** For very big x's or very small x's ($x = -1000, -2000$, etc), we can throw away all terms that do not involve the highest power of x; here, we throw away the 5; $\frac{2x}{3x} = \frac{2}{3}.$

C26: **C:** 149 total; the median is the grade of the 75th student, or 80.

C27: **D:** $x = 3$ and $y = \frac{1}{2}$, so $\frac{3x + 1}{y + 2} = \frac{10}{2.5} = 4.$

C28: **A:** Notice 2(5) = 10; with manipulations, all denominators are powers of 10! A = .06, B = .04, C = .012, D = .009 E = .0112.

C29: **B:** the others would imply $c = d = 0$. Also, we can rewrite as $c^2 - cd = 0$, which factors as $c(c-d) = 0$. Since $c \neq d$, c must equal 0.

C30: **E:** Kara will need 222 square inches. SA = $2(\ell \times w) + 2(h \times w) + 2(\ell \times h)$. Don't forget to change 1 foot to 12 inches.

C31: **E:** There are six: 1, 2, 3, 4, 5, and 0. Don't forget the 0!

C32: **E:** Each of the other answer choices is a side of the polygon. Choice A corresponds to AB, choice B corresponds to AD, choice C correponds to BC, and choice D corresponds to CD.

C33:   **C:** $(1)(2)(4) = 8$

C34:   **A:** $3^4 10^4 = 30^4$.

C35:   **A:** The sides are 10 and 2; trial and error works best for a quick answer. Algebraically, $L + W = 12$ and $LW = 20$. By substitution, $(W)(12 - W) = 20$ or $W^2 - 12W + 20 = 0$. Then $(W - 10)(W - 2) = 0$, which means $W = 10$ or $W = 2$. Select $W = 2$, the smaller number.

C36:   **C:** $\frac{1}{2}$ gallon $= 64$ ounces; 1 quart $= 32$ ounces; $1\frac{1}{2}$ pints $= (16 + 8)$ ounces, so the total is $(64 + 32 + 16 + 8) = 120$ ounces.

C37:   **C:** $F = C = \left(\frac{9}{5}\right)C + 32; \left(-\frac{4}{5}\right)C = 32; C = 32\left(-\frac{5}{4}\right) = 8(-5) = -40$.

Practice Test D

D1: In a parallel circuit in electricity, the total resistance R of resistors X and Y is given by the formula $\frac{1}{X} + \frac{1}{Y} = \frac{1}{R}$, where the resistance is in ohms. If $X = 12$ ohms and $Y = 6$ ohms, $R =$

   A.   3

   B.   4

   C.   8

   D.   18

   E.   24

D2: If $a = -2$ and $b = -3$, the value of $ab^2 - (ab)^2$ is

   A.   0

   B. -36

   C. -54

   D. -72

   E. -324

D3: $x^{12} =$

    A. $x^6 + x^6$

    B. $(x^6)^6$

    C. $\dfrac{x^3}{x^{15}}$

    D. $x^{17} - x^5$

    E. $x^3 x^9$

D4: If $\dfrac{x^2 - 3x - 18}{x + 4} = 0$, then $x =$

    A.   6 only

    B. -3 only

    C. -3 and 6 only

    D. $6, 0, -3$ only

    E. $-4, -3, 6$ only

D5: $\left(\sqrt{6} + \sqrt{6}\right)^2$

    A.   6

    B.   12

    C.   24

    D.   72

    E.   576

D6: $\dfrac{\frac{1}{3} + \frac{1}{4}}{\frac{1}{6}}$

    A. $\dfrac{1}{2}$

    B. $\dfrac{6}{7}$

    C. $\dfrac{12}{7}$

    D.   2.5

    E.   3.5

D7:   If the area of a square is p. In terms of p what is the diagonal?

A. $p\sqrt{2}$

B. $\sqrt{p}$

C. $\sqrt{2p}$

D. $2\sqrt{p}$

E. $\dfrac{p\sqrt{2}}{2}$

D8:   A monkey is trying to climb out of a 40-foot hole. Each morning he climbs up 4 feet. At night he gets so tired he falls back 3 feet. Every day this is repeated. On what day does he climb out of the hole?

A.   36

B.   37

C.   38

D.   39

E.   40

D9:   A 60-foot fence is in front of my house (not true), and it has a fence post every 6 feet. The number of fence posts is

A.   9

B.   10

C.   11

D.   12

E.   24

D10:   A 60-foot ribbon is cut into 6-foot pieces. The number of cuts is

A.   9

B.   10

C.   11

D.   12

E.   24

D11: 960 ounces of radioactive roast beef loses half its radioactivity every 4 hours. How many ounces of radioactive roast beef remain after one full day?

    A.   0

    B.   15

    C.   30

    D.   60

    E.   120

D12: If $n \neq 3$, then $\dfrac{2n^2(n-3) - n + 3}{n-3} =$

    A. $2n^2 - n + 3$

    B. $2n^2$

    C. $2n^2 - 1$

    D. $2n^2 + 1$

    E. $2n^2 + 3$

D13: $\dfrac{(.5)^8}{(.5)^4} =$

    A.   .25

    B.   .125

    C.   .0625

    D.   .03125

    E.   .015625

D14: A circle fits inside a square, touching all the sides of the square. If a side of the square is 6, what is the area of the circle?

    A. 3π

    B. 6π

    C. 9π

    D. $\sqrt{6\pi}$

    E.   not enough information is given

D15: If $4x + 3y = 11$ and $6x + 5y = 27$, the average (mean) of x and y is

A. 2

B. 4

C. 8

D. 9.5

E. 19

D16: A cube and a rectangular solid have the same volume. If the dimensions of the rectangular solid are 1, 3, and 9, how long is one side of the cube?

A. 1

B. 3

C. 13

D. $\sqrt[3]{13}$

E. 27

D17: If $M = \{$all even numbers$\}$ and $N = \{$all prime numbers$\}$, what is $M \cap N$?

A. $\{1, 3, 5, \ldots\}$

B. $\{2, 4, 6, \ldots\}$

C. $\{$all prime numbers$\}$

D. 2

E. ϕ

D18: If n is divided by 7, the remainder is 5. If $2n$ is divided by 7, the remainder is

A. 2

B. 3

C. 4

D. 5

E. 6

D19: If $(x - 2)^2 = 900$, the value of x could be

    A.   30

    B.   28

    C. -28

    D. -30

    E. -32

D20: A right triangle has sides 1 and $\sqrt{2}$. What can the third side be?

    I. 1

    II. $\sqrt{2}$

    III. $\sqrt{3}$

    A.   I only

    B.   II only

    C.   III only

    D.   I and II only

    E.   I and III only

D21: If $(x + y)^2 = 100$ and $xy = -3$, then $x^2 + y^2 =$

    A.   4

    B.   7

    C.   94

    D.   100

    E.   106

D22: Suppose we have an empty tank. One pipe fills the tank in 4 hours and a second pipe empties the tank in 6 hours. If the pipes are open together, how many hours will it take to fill the tank halfway?

    A.   2

    B.   4

    C.   6

    D.   8

    E.   12

D23: The number of boys is twice the number of girls. There are 60 children total. The pair of equations that best describes this is

  A. $b - 2g = 0, b + g = 60$

  B. $b + 2g = 0, b + g = 60$

  C. $b - 2g = 0, b + 2g = 60$

  D. $b + 2g = 0, 2b + g = 60$

  E. $2b - g = 0, b + g = 60.$

D24: If m and n are consecutive integers, which can never be even?

  A. mn

  B. $m + n$

  C. $n(m + 1)^2$

  D. $m(n - 1)^2$

  E. $m^2 + n^2 + 1$

D25: The price of a book is increased by 20%, and then the new price is discounted by 20%. Compared to the original price of the book, what is the final price?

  A.   10% less

  B.   4% less

  C.   Same as the original price

  D.   4% more

  E.   10% more

D26: The price of a second book is discounted 20%, and then the new price is increased by 20%. Compared to the original price of the book, what is the final price?

  A.   10% less

  B.   4% less

  C.   Same as the original price

  D.   4% more

  E.   10% more

D27: A $400 item is taxed at 5%. Then the whole price is discounted by 10%. A second $400 item is discounted at 10% and then the 5% sales tax is added on. The difference in cost between the two methods is

    A.   Nothing

    B. $10

    C. $20

    E. $30

    E. $40

D28: $\dfrac{31^2 + 31}{31} =$

    A.   31

    B.   32

    C.   62

    D.   941

    E.   942

D29: $3^6 + 3^6 + 3^6 =$

    A. 9^6

    B. 27^6

    C. 3^7

    D. 3^{18}

    E. 3^{216}

D30: If the diagonal of a square is 10, its area is

    A.   25

    B.   50

    C.   75

    D.   100

    E.   200

D31: If $4^{\frac{4}{3}}4^{\frac{5}{3}} = x^2$; x could equal

    A.   8

    B.   6

    C.   4

    D.   2

    E. -4

D32: If m books cost \$4 apiece and n books cost \$6 apiece, the mean cost per book is

    A. $\dfrac{24mn}{m+n}$

    B. $4m + \dfrac{6n}{mn}$

    C. $\dfrac{4m+6n}{m+n}$

    D. $\dfrac{4m+6n}{10}$

    E. $\dfrac{4m-6n}{m-n}$

D33: A taxi ride costs \$2.00 for the first quarter mile and 40 cents for each additional quarter of a mile. How much does a 3-mile trip cost?

    A. \$6.80

    B. \$6.40

    C. \$6.00

    D. \$5.60

    E. \$4.80

D34: $\dfrac{4}{1 + \frac{2}{x}} = 2; x =$

  A. 4

  B. 2

  C. 1

  D. $\dfrac{1}{2}$

  E. $\dfrac{1}{4}$

D35: A 6-sided die is tossed; then a coin is flipped. What is the probability that a head and a number greater than 4 will come up together?

  A. $\dfrac{1}{12}$

  B. $\dfrac{1}{6}$

  C. $\dfrac{1}{4}$

  D. $\dfrac{1}{3}$

  E. $\dfrac{5}{12}$

D36: What is the maximum number of $1\frac{1}{4}$-foot pieces that can be cut from a 32-foot board?

  A. 24

  B. 25

  C. 26

  D. 27

  E. 28

D37: How many cubic feet of dirt are found in a rectangular hole that is 6 feet long, 4 feet wide, and 2 inches deep?

    A. 48

    B. 24

    C. 6

    D. 4

    E. 0

(A) Let's look at the answers

D1: B: $\dfrac{1}{12} + \dfrac{1}{6} = \dfrac{1}{12} + \dfrac{2}{12} = \dfrac{3}{12} = \dfrac{1}{4} = \dfrac{1}{R}$; so $R = 4$.

D2: C: $(-2)(-3)^2 - ((-2)(-3))^2 = -18 - 36 = -54$

D3: E: $x^3 x^9 = x^{(3+9)} = x^{12}$

D4: C: For a fraction to equal 0, the top $= 0$ and the bottom $\neq 0$. So $x^2 - 3x - 18 = (x - 6)(x + 3) = 0$; and $x = 6$ and -3.

D5: C: $(\sqrt{6} + \sqrt{6})^2 = (2\sqrt{6})^2 = 4(6) = 24$.

D6: E: Multiply top and bottom by 12, and you get $\dfrac{7}{2}$, or 3.5.

D7: C: Each side is $s = \sqrt{p}$. The diagonal is then $s\sqrt{2} = \sqrt{p}\sqrt{2} = \sqrt{2p}$.

D8: B: After 36 days, the monkey is 36 feet up. On the 37th day, the monkey climbs up 4 feet. It is now at 40 feet and out of the hole. It does not fall back.

D9: C: $\dfrac{60}{6} = 10$. Then add 1 for the post at the beginning.

D10: A: You need only 9 cuts to get 10 pieces.

D11: B: 480 left after 4 hours; 240 left after 8 hours; 120 left after 12 hours; 60 left after 16 hours; 30 left after 20 hours; and 15 left after 24 hours $= 1$ day.

D12: C: $\dfrac{2n^2(n-3) - n + 3}{n - 3} = \dfrac{2n^2(n-3) - 1(n-3)}{(n-3)} = \dfrac{(n-3)(2n^2 - 1)}{(n-3)} = 2n^2 - 1$.

D13: C: $(.5)^4 = \left(\dfrac{1}{2}\right)^4 = \dfrac{1}{16} = .0625$.

D14:   **C:**   The diameter of the circle must be the same as the side of the square, which is 6. So $r = 3$, and the area is $\pi r^2 = \pi(3^2) = 9\pi$.

D15:   **B:**   Subtracting the first equation from the second, we get $2x + 2y = 16$; $x + y = 8$; so $\dfrac{x + y}{2} = 4$.

D16:   **B:**   The volume of the rectangular solid is $V = \ell wh = 27$; the volume of a cube is s^3. Then the length of a side of the cube is $\sqrt[3]{27} = 3$

D17:   **D:**   The only even prime number is 2.

D18:   **B:**   Doubling the number doubles the remainder, so the remainder is 10, but $\dfrac{10}{7}$ has a remainder of 3.

D19:   **C:** $x - 2 = \pm 30$; so $x = 2 + 30 = 32$ or $2 - 30 = -28$.

D20:   **E:**   1 is okay since $(1)^2 + (1)^2 = \left(\sqrt{2}\right)^2$, and $\sqrt{3}$ is okay since $1^2 + \left(\sqrt{2}\right)^2 = \left(\sqrt{3}\right)^2$.

D21:   **E:** $x^2 + 2xy + y^2 = 100$; $x^2 + y^2 + 2(-3) = 100$; $x^2 + y^2 = 106$.

D22:   **C:** $\dfrac{x}{4} - \dfrac{x}{6} = \dfrac{1}{2}$; $3x - 2x = 6$; $x = 6$.

D23:   **A:**   The number of boys is twice the number of girls can be expressed as $b = 2g$, which is equivalent to $b - 2g = 0$. Since there are 60 children, the number of boys added to the number of girls is 60. This is written algebraically as $b + g = 60$.

D24:   **B:**   An odd plus an even always equals an odd.

D25:   **B:** $100 increased by 20% = $120; 20% off is $24; $120 − $24 = $96, which is 4% less than the original price.

D26:   **B:**   20% off $100 is $80; 20% of $80 is $16. $80 + $16 = $96, which is 4% less than the original price.

D27:   **A:**   Each gives a final price of $378.

D28:   **B:** $31(31 + 1)/31 = 31 + 1 = 32$.

D29:   **C:**   The sum is $3(3^6) = 3^1 3^6 = 3^7$.

D30:   **B:**   The easy formula to remember is $A = \left(\dfrac{1}{2}\right) d^2$; or $\left(\dfrac{10}{\sqrt{2}}\right)^2 = \dfrac{100}{2} = 50$.

D31:   **A:** $4^3 = x^2$; $x^2 = 64$; $x = \pm 8$.

D32: **C:** $\text{Mean} = \dfrac{\text{Total cost of books}}{\text{Total number of books}} = \dfrac{4m + 6n}{m + n}$.

D33: **B:** After the first quarter mile, there are 11 quarter miles left. The cost is $2.00 + 11(.40) = \$6.40$.

D34: **B:** 2. Try to do it by sight; $\dfrac{2}{x}$ must equal 1. Another approach is to multiply each side of the equation by $1 + \dfrac{2}{x}$ to get $4 = 2\left(1 + \dfrac{2}{x}\right)$, which simplifies to $4 = 2 + \dfrac{4}{x}$.

Then, $2 = \dfrac{4}{x}$, so $x = 2$.

D35: **B:** $\text{Probability} = \dfrac{\text{Success}}{\text{Possibilities}}$. The probabilty of a 5 or 6 on a 6-sided die is $\dfrac{2}{6}$, and the probability is $\dfrac{1}{2}$ for a head on a coin. $\dfrac{2}{6} \times \dfrac{1}{2} = \dfrac{1}{6}$, since $31\dfrac{1}{4} + 1\dfrac{1}{4} > 32$

D36: **B:** There will be 4 pieces every 5 feet; or 24 pieces for 30 feet; or 25 pieces for $31\dfrac{1}{4}$ feet. The next piece will be $1\dfrac{1}{4}$ feet; the answer is 25, since $31\dfrac{1}{4} + 1\dfrac{1}{4} > 32$.

D37: **E:** The last question in the book is an old joke. The answer is there is no dirt in a hole. The point to this question, as with many questions on this test, is to read all questions very carefully and as quickly as possible. If you misread the question, no matter how good your math is, you will miss it.

CHAPTER 17: *Epilogue*

"*If you feel more is needed, then you may need a little more for your confidence. But if our journey together has been complete, you have all you need for success.*"

Again, congratulations. Now that you've finished the book, you might do several things. First, review any section that has caused you difficulty. Next, get more questions if you need to. You can get the official book published by the Graduate Management Admission Council. If you still need more questions, Educational Testing Service (ETS), which used to control this test, has an older book with questions. Finally, you might contact the GMAC to get sample computer problems to try.

Remember again that you do not need a perfect score. You only need a score that will get you into the graduate school of your choice.

Good luck with the rest of your college education and the rest of your life.

ANSWER SHEET: Data Sufficiency - *Practice Test A*

1. Ⓐ Ⓑ Ⓒ Ⓓ Ⓔ

2. Ⓐ Ⓑ Ⓒ Ⓓ Ⓔ

3. Ⓐ Ⓑ Ⓒ Ⓓ Ⓔ

4. Ⓐ Ⓑ Ⓒ Ⓓ Ⓔ

5. Ⓐ Ⓑ Ⓒ Ⓓ Ⓔ

6. Ⓐ Ⓑ Ⓒ Ⓓ Ⓔ

7. Ⓐ Ⓑ Ⓒ Ⓓ Ⓔ

8. Ⓐ Ⓑ Ⓒ Ⓓ Ⓔ

9. Ⓐ Ⓑ Ⓒ Ⓓ Ⓔ

10. Ⓐ Ⓑ Ⓒ Ⓓ Ⓔ

11. Ⓐ Ⓑ Ⓒ Ⓓ Ⓔ

12. Ⓐ Ⓑ Ⓒ Ⓓ Ⓔ

13. Ⓐ Ⓑ Ⓒ Ⓓ Ⓔ

14. Ⓐ Ⓑ Ⓒ Ⓓ Ⓔ

15. Ⓐ Ⓑ Ⓒ Ⓓ Ⓔ

16. Ⓐ Ⓑ Ⓒ Ⓓ Ⓔ

17. Ⓐ Ⓑ Ⓒ Ⓓ Ⓔ

18. Ⓐ Ⓑ Ⓒ Ⓓ Ⓔ

19. Ⓐ Ⓑ Ⓒ Ⓓ Ⓔ

20. Ⓐ Ⓑ Ⓒ Ⓓ Ⓔ

21. Ⓐ Ⓑ Ⓒ Ⓓ Ⓔ

22. Ⓐ Ⓑ Ⓒ Ⓓ Ⓔ

23. Ⓐ Ⓑ Ⓒ Ⓓ Ⓔ

24. Ⓐ Ⓑ Ⓒ Ⓓ Ⓔ

25. Ⓐ Ⓑ Ⓒ Ⓓ Ⓔ

26. Ⓐ Ⓑ Ⓒ Ⓓ Ⓔ

27. Ⓐ Ⓑ Ⓒ Ⓓ Ⓔ

28. Ⓐ Ⓑ Ⓒ Ⓓ Ⓔ

29. Ⓐ Ⓑ Ⓒ Ⓓ Ⓔ

30. Ⓐ Ⓑ Ⓒ Ⓓ Ⓔ

31. Ⓐ Ⓑ Ⓒ Ⓓ Ⓔ

32. Ⓐ Ⓑ Ⓒ Ⓓ Ⓔ

33. Ⓐ Ⓑ Ⓒ Ⓓ Ⓔ

34. Ⓐ Ⓑ Ⓒ Ⓓ Ⓔ

35. Ⓐ Ⓑ Ⓒ Ⓓ Ⓔ

36. Ⓐ Ⓑ Ⓒ Ⓓ Ⓔ

37. Ⓐ Ⓑ Ⓒ Ⓓ Ⓔ

ANSWER SHEET: Data Sufficiency - *Practice Test B*

1. Ⓐ Ⓑ Ⓒ Ⓓ Ⓔ
2. Ⓐ Ⓑ Ⓒ Ⓓ Ⓔ
3. Ⓐ Ⓑ Ⓒ Ⓓ Ⓔ
4. Ⓐ Ⓑ Ⓒ Ⓓ Ⓔ
5. Ⓐ Ⓑ Ⓒ Ⓓ Ⓔ
6. Ⓐ Ⓑ Ⓒ Ⓓ Ⓔ
7. Ⓐ Ⓑ Ⓒ Ⓓ Ⓔ
8. Ⓐ Ⓑ Ⓒ Ⓓ Ⓔ
9. Ⓐ Ⓑ Ⓒ Ⓓ Ⓔ
10. Ⓐ Ⓑ Ⓒ Ⓓ Ⓔ
11. Ⓐ Ⓑ Ⓒ Ⓓ Ⓔ
12. Ⓐ Ⓑ Ⓒ Ⓓ Ⓔ
13. Ⓐ Ⓑ Ⓒ Ⓓ Ⓔ
14. Ⓐ Ⓑ Ⓒ Ⓓ Ⓔ
15. Ⓐ Ⓑ Ⓒ Ⓓ Ⓔ
16. Ⓐ Ⓑ Ⓒ Ⓓ Ⓔ
17. Ⓐ Ⓑ Ⓒ Ⓓ Ⓔ
18. Ⓐ Ⓑ Ⓒ Ⓓ Ⓔ
19. Ⓐ Ⓑ Ⓒ Ⓓ Ⓔ

20. Ⓐ Ⓑ Ⓒ Ⓓ Ⓔ
21. Ⓐ Ⓑ Ⓒ Ⓓ Ⓔ
22. Ⓐ Ⓑ Ⓒ Ⓓ Ⓔ
23. Ⓐ Ⓑ Ⓒ Ⓓ Ⓔ
24. Ⓐ Ⓑ Ⓒ Ⓓ Ⓔ
25. Ⓐ Ⓑ Ⓒ Ⓓ Ⓔ
26. Ⓐ Ⓑ Ⓒ Ⓓ Ⓔ
27. Ⓐ Ⓑ Ⓒ Ⓓ Ⓔ
28. Ⓐ Ⓑ Ⓒ Ⓓ Ⓔ
29. Ⓐ Ⓑ Ⓒ Ⓓ Ⓔ
30. Ⓐ Ⓑ Ⓒ Ⓓ Ⓔ
31. Ⓐ Ⓑ Ⓒ Ⓓ Ⓔ
32. Ⓐ Ⓑ Ⓒ Ⓓ Ⓔ
33. Ⓐ Ⓑ Ⓒ Ⓓ Ⓔ
34. Ⓐ Ⓑ Ⓒ Ⓓ Ⓔ
35. Ⓐ Ⓑ Ⓒ Ⓓ Ⓔ
36. Ⓐ Ⓑ Ⓒ Ⓓ Ⓔ
37. Ⓐ Ⓑ Ⓒ Ⓓ Ⓔ

ANSWER SHEET: Data Sufficiency - *Practice Test C*

1. Ⓐ Ⓑ Ⓒ Ⓓ Ⓔ

2. Ⓐ Ⓑ Ⓒ Ⓓ Ⓔ

3. Ⓐ Ⓑ Ⓒ Ⓓ Ⓔ

4. Ⓐ Ⓑ Ⓒ Ⓓ Ⓔ

5. Ⓐ Ⓑ Ⓒ Ⓓ Ⓔ

6. Ⓐ Ⓑ Ⓒ Ⓓ Ⓔ

7. Ⓐ Ⓑ Ⓒ Ⓓ Ⓔ

8. Ⓐ Ⓑ Ⓒ Ⓓ Ⓔ

9. Ⓐ Ⓑ Ⓒ Ⓓ Ⓔ

10. Ⓐ Ⓑ Ⓒ Ⓓ Ⓔ

11. Ⓐ Ⓑ Ⓒ Ⓓ Ⓔ

12. Ⓐ Ⓑ Ⓒ Ⓓ Ⓔ

13. Ⓐ Ⓑ Ⓒ Ⓓ Ⓔ

14. Ⓐ Ⓑ Ⓒ Ⓓ Ⓔ

15. Ⓐ Ⓑ Ⓒ Ⓓ Ⓔ

16. Ⓐ Ⓑ Ⓒ Ⓓ Ⓔ

17. Ⓐ Ⓑ Ⓒ Ⓓ Ⓔ

18. Ⓐ Ⓑ Ⓒ Ⓓ Ⓔ

19. Ⓐ Ⓑ Ⓒ Ⓓ Ⓔ

20. Ⓐ Ⓑ Ⓒ Ⓓ Ⓔ

21. Ⓐ Ⓑ Ⓒ Ⓓ Ⓔ

22. Ⓐ Ⓑ Ⓒ Ⓓ Ⓔ

23. Ⓐ Ⓑ Ⓒ Ⓓ Ⓔ

24. Ⓐ Ⓑ Ⓒ Ⓓ Ⓔ

25. Ⓐ Ⓑ Ⓒ Ⓓ Ⓔ

26. Ⓐ Ⓑ Ⓒ Ⓓ Ⓔ

27. Ⓐ Ⓑ Ⓒ Ⓓ Ⓔ

28. Ⓐ Ⓑ Ⓒ Ⓓ Ⓔ

29. Ⓐ Ⓑ Ⓒ Ⓓ Ⓔ

30. Ⓐ Ⓑ Ⓒ Ⓓ Ⓔ

31. Ⓐ Ⓑ Ⓒ Ⓓ Ⓔ

32. Ⓐ Ⓑ Ⓒ Ⓓ Ⓔ

33. Ⓐ Ⓑ Ⓒ Ⓓ Ⓔ

34. Ⓐ Ⓑ Ⓒ Ⓓ Ⓔ

35. Ⓐ Ⓑ Ⓒ Ⓓ Ⓔ

36. Ⓐ Ⓑ Ⓒ Ⓓ Ⓔ

37. Ⓐ Ⓑ Ⓒ Ⓓ Ⓔ

ANSWER SHEET: Problem Solving - *Practice Test A*

1. Ⓐ Ⓑ Ⓒ Ⓓ Ⓔ

2. Ⓐ Ⓑ Ⓒ Ⓓ Ⓔ

3. Ⓐ Ⓑ Ⓒ Ⓓ Ⓔ

4. Ⓐ Ⓑ Ⓒ Ⓓ Ⓔ

5. Ⓐ Ⓑ Ⓒ Ⓓ Ⓔ

6. Ⓐ Ⓑ Ⓒ Ⓓ Ⓔ

7. Ⓐ Ⓑ Ⓒ Ⓓ Ⓔ

8. Ⓐ Ⓑ Ⓒ Ⓓ Ⓔ

9. Ⓐ Ⓑ Ⓒ Ⓓ Ⓔ

10. Ⓐ Ⓑ Ⓒ Ⓓ Ⓔ

11. Ⓐ Ⓑ Ⓒ Ⓓ Ⓔ

12. Ⓐ Ⓑ Ⓒ Ⓓ Ⓔ

13. Ⓐ Ⓑ Ⓒ Ⓓ Ⓔ

14. Ⓐ Ⓑ Ⓒ Ⓓ Ⓔ

15. Ⓐ Ⓑ Ⓒ Ⓓ Ⓔ

16. Ⓐ Ⓑ Ⓒ Ⓓ Ⓔ

17. Ⓐ Ⓑ Ⓒ Ⓓ Ⓔ

18. Ⓐ Ⓑ Ⓒ Ⓓ Ⓔ

19. Ⓐ Ⓑ Ⓒ Ⓓ Ⓔ

20. Ⓐ Ⓑ Ⓒ Ⓓ Ⓔ

21. Ⓐ Ⓑ Ⓒ Ⓓ Ⓔ

22. Ⓐ Ⓑ Ⓒ Ⓓ Ⓔ

23. Ⓐ Ⓑ Ⓒ Ⓓ Ⓔ

24. Ⓐ Ⓑ Ⓒ Ⓓ Ⓔ

25. Ⓐ Ⓑ Ⓒ Ⓓ Ⓔ

26. Ⓐ Ⓑ Ⓒ Ⓓ Ⓔ

27. Ⓐ Ⓑ Ⓒ Ⓓ Ⓔ

28. Ⓐ Ⓑ Ⓒ Ⓓ Ⓔ

29. Ⓐ Ⓑ Ⓒ Ⓓ Ⓔ

30. Ⓐ Ⓑ Ⓒ Ⓓ Ⓔ

31. Ⓐ Ⓑ Ⓒ Ⓓ Ⓔ

32. Ⓐ Ⓑ Ⓒ Ⓓ Ⓔ

33. Ⓐ Ⓑ Ⓒ Ⓓ Ⓔ

34. Ⓐ Ⓑ Ⓒ Ⓓ Ⓔ

35. Ⓐ Ⓑ Ⓒ Ⓓ Ⓔ

36. Ⓐ Ⓑ Ⓒ Ⓓ Ⓔ

37. Ⓐ Ⓑ Ⓒ Ⓓ Ⓔ

ANSWER SHEET: Problem Solving - *Practice Test B*

1. Ⓐ Ⓑ Ⓒ Ⓓ Ⓔ

2. Ⓐ Ⓑ Ⓒ Ⓓ Ⓔ

3. Ⓐ Ⓑ Ⓒ Ⓓ Ⓔ

4. Ⓐ Ⓑ Ⓒ Ⓓ Ⓔ

5. Ⓐ Ⓑ Ⓒ Ⓓ Ⓔ

6. Ⓐ Ⓑ Ⓒ Ⓓ Ⓔ

7. Ⓐ Ⓑ Ⓒ Ⓓ Ⓔ

8. Ⓐ Ⓑ Ⓒ Ⓓ Ⓔ

9. Ⓐ Ⓑ Ⓒ Ⓓ Ⓔ

10. Ⓐ Ⓑ Ⓒ Ⓓ Ⓔ

11. Ⓐ Ⓑ Ⓒ Ⓓ Ⓔ

12. Ⓐ Ⓑ Ⓒ Ⓓ Ⓔ

13. Ⓐ Ⓑ Ⓒ Ⓓ Ⓔ

14. Ⓐ Ⓑ Ⓒ Ⓓ Ⓔ

15. Ⓐ Ⓑ Ⓒ Ⓓ Ⓔ

16. Ⓐ Ⓑ Ⓒ Ⓓ Ⓔ

17. Ⓐ Ⓑ Ⓒ Ⓓ Ⓔ

18. Ⓐ Ⓑ Ⓒ Ⓓ Ⓔ

19. Ⓐ Ⓑ Ⓒ Ⓓ Ⓔ

20. Ⓐ Ⓑ Ⓒ Ⓓ Ⓔ

21. Ⓐ Ⓑ Ⓒ Ⓓ Ⓔ

22. Ⓐ Ⓑ Ⓒ Ⓓ Ⓔ

23. Ⓐ Ⓑ Ⓒ Ⓓ Ⓔ

24. Ⓐ Ⓑ Ⓒ Ⓓ Ⓔ

25. Ⓐ Ⓑ Ⓒ Ⓓ Ⓔ

26. Ⓐ Ⓑ Ⓒ Ⓓ Ⓔ

27. Ⓐ Ⓑ Ⓒ Ⓓ Ⓔ

28. Ⓐ Ⓑ Ⓒ Ⓓ Ⓔ

29. Ⓐ Ⓑ Ⓒ Ⓓ Ⓔ

30. Ⓐ Ⓑ Ⓒ Ⓓ Ⓔ

31. Ⓐ Ⓑ Ⓒ Ⓓ Ⓔ

32. Ⓐ Ⓑ Ⓒ Ⓓ Ⓔ

33. Ⓐ Ⓑ Ⓒ Ⓓ Ⓔ

34. Ⓐ Ⓑ Ⓒ Ⓓ Ⓔ

35. Ⓐ Ⓑ Ⓒ Ⓓ Ⓔ

36. Ⓐ Ⓑ Ⓒ Ⓓ Ⓔ

37. Ⓐ Ⓑ Ⓒ Ⓓ Ⓔ

ANSWER SHEET: Problem Solving - *Practice Test C*

1. Ⓐ Ⓑ Ⓒ Ⓓ Ⓔ

2. Ⓐ Ⓑ Ⓒ Ⓓ Ⓔ

3. Ⓐ Ⓑ Ⓒ Ⓓ Ⓔ

4. Ⓐ Ⓑ Ⓒ Ⓓ Ⓔ

5. Ⓐ Ⓑ Ⓒ Ⓓ Ⓔ

6. Ⓐ Ⓑ Ⓒ Ⓓ Ⓔ

7. Ⓐ Ⓑ Ⓒ Ⓓ Ⓔ

8. Ⓐ Ⓑ Ⓒ Ⓓ Ⓔ

9. Ⓐ Ⓑ Ⓒ Ⓓ Ⓔ

10. Ⓐ Ⓑ Ⓒ Ⓓ Ⓔ

11. Ⓐ Ⓑ Ⓒ Ⓓ Ⓔ

12. Ⓐ Ⓑ Ⓒ Ⓓ Ⓔ

13. Ⓐ Ⓑ Ⓒ Ⓓ Ⓔ

14. Ⓐ Ⓑ Ⓒ Ⓓ Ⓔ

15. Ⓐ Ⓑ Ⓒ Ⓓ Ⓔ

16. Ⓐ Ⓑ Ⓒ Ⓓ Ⓔ

17. Ⓐ Ⓑ Ⓒ Ⓓ Ⓔ

18. Ⓐ Ⓑ Ⓒ Ⓓ Ⓔ

19. Ⓐ Ⓑ Ⓒ Ⓓ Ⓔ

20. Ⓐ Ⓑ Ⓒ Ⓓ Ⓔ

21. Ⓐ Ⓑ Ⓒ Ⓓ Ⓔ

22. Ⓐ Ⓑ Ⓒ Ⓓ Ⓔ

23. Ⓐ Ⓑ Ⓒ Ⓓ Ⓔ

24. Ⓐ Ⓑ Ⓒ Ⓓ Ⓔ

25. Ⓐ Ⓑ Ⓒ Ⓓ Ⓔ

26. Ⓐ Ⓑ Ⓒ Ⓓ Ⓔ

27. Ⓐ Ⓑ Ⓒ Ⓓ Ⓔ

28. Ⓐ Ⓑ Ⓒ Ⓓ Ⓔ

29. Ⓐ Ⓑ Ⓒ Ⓓ Ⓔ

30. Ⓐ Ⓑ Ⓒ Ⓓ Ⓔ

31. Ⓐ Ⓑ Ⓒ Ⓓ Ⓔ

32. Ⓐ Ⓑ Ⓒ Ⓓ Ⓔ

33. Ⓐ Ⓑ Ⓒ Ⓓ Ⓔ

34. Ⓐ Ⓑ Ⓒ Ⓓ Ⓔ

35. Ⓐ Ⓑ Ⓒ Ⓓ Ⓔ

36. Ⓐ Ⓑ Ⓒ Ⓓ Ⓔ

37. Ⓐ Ⓑ Ⓒ Ⓓ Ⓔ

ANSWER SHEET: Problem Solving - *Practice Test D*

1. Ⓐ Ⓑ Ⓒ Ⓓ Ⓔ

2. Ⓐ Ⓑ Ⓒ Ⓓ Ⓔ

3. Ⓐ Ⓑ Ⓒ Ⓓ Ⓔ

4. Ⓐ Ⓑ Ⓒ Ⓓ Ⓔ

5. Ⓐ Ⓑ Ⓒ Ⓓ Ⓔ

6. Ⓐ Ⓑ Ⓒ Ⓓ Ⓔ

7. Ⓐ Ⓑ Ⓒ Ⓓ Ⓔ

8. Ⓐ Ⓑ Ⓒ Ⓓ Ⓔ

9. Ⓐ Ⓑ Ⓒ Ⓓ Ⓔ

10. Ⓐ Ⓑ Ⓒ Ⓓ Ⓔ

11. Ⓐ Ⓑ Ⓒ Ⓓ Ⓔ

12. Ⓐ Ⓑ Ⓒ Ⓓ Ⓔ

13. Ⓐ Ⓑ Ⓒ Ⓓ Ⓔ

14. Ⓐ Ⓑ Ⓒ Ⓓ Ⓔ

15. Ⓐ Ⓑ Ⓒ Ⓓ Ⓔ

16. Ⓐ Ⓑ Ⓒ Ⓓ Ⓔ

17. Ⓐ Ⓑ Ⓒ Ⓓ Ⓔ

18. Ⓐ Ⓑ Ⓒ Ⓓ Ⓔ

19. Ⓐ Ⓑ Ⓒ Ⓓ Ⓔ

20. Ⓐ Ⓑ Ⓒ Ⓓ Ⓔ

21. Ⓐ Ⓑ Ⓒ Ⓓ Ⓔ

22. Ⓐ Ⓑ Ⓒ Ⓓ Ⓔ

23. Ⓐ Ⓑ Ⓒ Ⓓ Ⓔ

24. Ⓐ Ⓑ Ⓒ Ⓓ Ⓔ

25. Ⓐ Ⓑ Ⓒ Ⓓ Ⓔ

26. Ⓐ Ⓑ Ⓒ Ⓓ Ⓔ

27. Ⓐ Ⓑ Ⓒ Ⓓ Ⓔ

28. Ⓐ Ⓑ Ⓒ Ⓓ Ⓔ

29. Ⓐ Ⓑ Ⓒ Ⓓ Ⓔ

30. Ⓐ Ⓑ Ⓒ Ⓓ Ⓔ

31. Ⓐ Ⓑ Ⓒ Ⓓ Ⓔ

32. Ⓐ Ⓑ Ⓒ Ⓓ Ⓔ

33. Ⓐ Ⓑ Ⓒ Ⓓ Ⓔ

34. Ⓐ Ⓑ Ⓒ Ⓓ Ⓔ

35. Ⓐ Ⓑ Ⓒ Ⓓ Ⓔ

36. Ⓐ Ⓑ Ⓒ Ⓓ Ⓔ

37. Ⓐ Ⓑ Ⓒ Ⓓ Ⓔ

INDEX